God, Money, and Politics

English Attitudes to Blindness and Touch, from the Enlightenment to Integration

God, Money, and Politics

English Attitudes to Blindness and Touch, from the Enlightenment to Integration

by

Simon Hayhoe

IAP

INFORMATION AGE PUBLISHING, INC.
Charlotte, NC • www.infoagepub.com

Library of Congress Cataloging-in-Publication Data

Hayhoe, Simon.
God, money, and politics : English attitudes to blindness and touch, from the Enlightenment to integration / by Simon Hayhoe.
p. cm.
Includes bibliographical references.
ISBN 978-1-59311-913-3 (pbk.) – ISBN 978-1-59311-914-0 (hardcover)
1. Blind–Great Britain. 2. Blind–Education–Great Britain–History. 3. Great Britain. Education Act 1981. 4. Blindness. 5. Touch. I. Title.
HV1945.H39 2008
362.4'10942–dc22

2008008676

Printed and bound by CPI Group (UK) Ltd, Croydon, CR0 4YY

for Vanessa

CONTENTS

PART I

PART II

PART III

ACKNOWLEDGMENTS

The author would like to acknowledge the following for their help in the writing of this book:

- Duke Humfrey's Library, Bodleian Library, Oxford University, UK
- Manuscripts Collection, Unviersity Library, Cambridge University, UK
- The Research Library, Royal National Institute for the Blind, UK
- Bristol Royal Society for the Blind, UK
- School of Education Library, Birmingham University, UK
- Perkins Research Library, Perkins College for the Blind, Mass, US

CHAPTER 1

INTRODUCTION: FROM ENLIGHTENMENT TO INTEGRATION

> A man being born blind and having a globe and a cube, nigh of the same bigness, committed into his hand, and being taught or told, which is called the globe and which the cube, so as easily to distinguish them by touch or feeling; then both things taken from him, and laid on a table. Let us suppose his sight restored to him; whether he could, by his sight, and before he touch them, know which is the globe and which the cube? So whether he could not reach them though they were removed 20 or 1000 feet from him?
>
> [MS, Locke c16 ff. 92–93 Letter from Molyneux (William) to the authors of the *Bibliotheque universelle,* 7 July 1688. Reproduced from the correspondence of John Locke, the John Locke Collection, Bodleian Library, Oxford University]

The late seventeenth century—after the publication of William Molyneux's question to John Locke—and the late twentieth century—when the first true legal step to integration of disability in English society, the 1981 Education Act, was passed—began two important eras of revolutionary change in attitudes to blindness and touch in England. The intervening years also led us to question the accepted view that disability is a relatively homogeneous notion (Barnes & Mercer, 2003), rather than a collection of diverse, unique histories. The changes that these two years achieved sewed the seeds of different types of institutions for the blind, concerned not just with practical and academic education, but socialization and inculcation into a culture

God, Money, and Politics, pages 1–13

that had previously rejected them; one type physical the other virtual; one type would help to socially cleanse a society, deny the blind population intellectual education, and provide their purifiers an exalted status; the other type would set an agenda for closing these institutions, help their administrators gain political power, and leave a blind population to survive in mainstream schools unaided. In the history of these two years, however, there are distinct similarities in their institutions' political and cultural environments.

To begin with, both eras were set in a time of national turmoil and a shift in the power of state institutions from one ideological administration to another. At the same time there developed an increasingly homogenic development of cultural thinking in broader society. Both periods were also set against a backdrop of English-speaking philosophy becoming the engine of thought in the Western world. In England, Wales, Ireland, and Scotland in the 1690s, this pursuit of academic power was undergoing rhetorically separate yet culturally related social changes of a fundamental nature. This change occurred because, for the most part, these countries' common heads of state and religious trends would shape their societies for years to come. Less than twenty years after Molyneux's letter—in 1707—what would be described as a British culture would be founded, prior to a revolution of Empire, industry and a worldwide capital production fired by the industrial revolution. In the sixteenth, seventeenth, and eighteenth centuries, England also held itself up as one of the protectors of the Protestant faith in Western Europe. From the eighteenth century, the increase in Protestant evangelism and the growth of colonization also led to the procreation of these Christian models as world religions.

Moreover, in the England of the seventeenth century religion was still of fundamental importance to the political, academic and institutional structure. The 1690s were part of an era in which England was being torn between Roman Catholic and Protestant influenced politics, and the theocratic orthodoxies of the presiding link between Church and State were being questioned by radical theologians, such as the Latitudinarians. Molyneux and Locke also corresponded in a period of English history when the whole nature of political administration was under question. It was the end of a century when there had been a bloody civil war, when there was still an unsettled atmosphere of Monarchism and Protectorism (the formation of a republic in what was to become Britain), where the House of Commons finally became more powerful than the monarch and the House of Lords—when, there can be little doubt, the roots of the current English political ethic were first nurtured.

By contrast the 1980s, as England approached the end of the twentieth century, its culture was increasingly influenced by the use of money as a means to pursue power. Moreover, social class was being defined less by heritage, such as title and peerage, or following the profession of your father, or geographic location, but was defined more by how much money you earned, what school, style of clothes, household equipment or car you could afford, or the area or retailer where you could afford them. After a century of two world wars, Britain had also found itself in a revolution of alternative politics, in national debates on issues such as gender and race. People became *liberated* from traditional tasks. As men were called to fight on the front—this was also an era when their government made them join a single national armed force—women filled their roles at home.

After the Second World War, Britain's failing economy needed extra workers as it restructured its infrastructure, broken irrevocably by collateral damage and, more important, the death and maiming of so many men. This began a wave of immigration into the British Isles from its current and former colonies, from the mid 1950s to the late 1970s and was coupled with an increase in the number of women in the work place. The end of the 1970s, however, also saw an increasing financial crisis, national strikes and a swell of resentment of existing politics by a large proportion of mainly older working class people and an ever increasing under class. These feelings were to lead in turn to an ever more fractious social system and welfare state. Such political conflict was also fueled by a growing divide between new middle class professionals and people from the working and under classes, who increasingly found themselves unemployed—this period also saw the beginning of a drastic increase in middle class women attending university. Most important, the middle classes had a vastly greater access to information and good jobs, and at the end of the decade were supported by a government that from the late 1970s was led by Margaret Thatcher; when industry started drowning, they found it easier to survive.

The world wars had also led to a change in the educational structure of England in the twentieth century. Although there had been widespread public education in England since the nineteenth century, the war had encouraged a more liberal education, particularly for adults (Hayhoe, 1995a) and particularly those in the armed forces. With the introduction of National Service during the Second World War, different social classes of soldiers found themselves in the same sleeping huts and the same ranks. They also had the experience of fighting together in teams as never before, before returning to their civilian lives and their divided social roles.

In addition, groups such as the Workers Educational Association not only introduced liberal arts courses for the troops, they encouraged these men to question their positions in society and the right of social privilege. They were also taught about the economic structures of broader society, and questioned the role of the class system. For children, too, the latter half of the century was characterized by a *changing* of state education, which confused the social roles of the traditional class system. These activities, coupled with the growth and organization of the union movement in previous decades, led to the radicalization of social thinking as never before, and demands for greater wages and status through material accumulation of manual and professional workers alike.

In addition to these social revolutions of economic attainment and power, the world in the twentieth century also underwent a substantial change in its communication structures, leading to further radical social theories and revolutions. The invention of easily reproducible photographs toward the end of the nineteenth century had been followed by the easy reproduction and printing of still and moving images at the beginning of the twentieth century. In the following decades, the inventions of radio, the telephone and then television, digital technology and computation made it possible to hear and see world events in far flung places almost instantly. In the 1970s the term The Global Village was coined, which increased the spread of international political movements with homogeneous aims, such as CND, Greenpeace and disability rights groups. All of this, coupled with the widespread distribution of new media and art forms, led to the usurpation of local, traditional cultures by a global musical, literary and visual culture.

Within this media revolution, English became the dominant language and a new international, intellectual middle class emerged, drawing in large part from Anglo-Saxon culture. This formed an increasingly *new world culture*, which grew to such an extent that by the late 1960s the Vietnam War was called the first television war, and the Beatles were in most homes in the Western world. Such graphic images alongside the increased pace of social change occurring in the latter half of the twentieth century led many more of the population to question the nature of social structures and activities that had previously been taken for granted.

So how did these social changes affect our notions of blindness and touch?

Our book sets out on a discussion of this question, and in doing so examines the role of three factors, God, Money, and Politics, in what we will call the epistemological theory of blindness; that is the theory of the construction of knowledge on blindness and touch by social and cultural change (Hayhoe, 2006). This book also illustrates how this development has, in the main, been motivated by an attempt to assert or gain power. Before activation of this task, however, it is first crucial to outline why the study of blindness in conventional academic subjects such as psychology, history and sociology is so important. We will do this by presenting the main theories of disability and blindness that have informed the writing of this book, and in doing so create a frame of reference for the presentation of the historical story that is written in the following chapters. This is essential in order to place the book in the broad context of these theories of disability and blindness. Such an exercise is also important as it is imperative to understand that blindness exists within an academic and symbolic context of physical impairment and the social mythologies that accompany such understanding.

INFLUENTIAL STUDIES ON THE UNDERSTANDING OF BLINDNESS AND DISABILITY

Although not as prevalent as many other physical or mental disabilities, blindness has been the particular subject of early institutional theories on disability, and of philosophies of interaction and behavior in the study of disability. Of particular interest to these thinkers is the question of whether it is possible for blind people to obtain as much information about the world through their remaining senses as sighted people through vision. In many situations this question still remains unanswered, as nonvisual perception is too complex a phenomenon to compare directly to sight, and perception differs in diverse circumstances. Moreover, such studies also derive problems because they assume that there is a group of people called the blind, who share a range of non visual experiences. Similarly, such discussions must assume a common visual experience, and a form of perception devoid of other sensual influences.

In the study of aesthetics, for example, the discussion about the mental ability of people who are blind has also been stretched beyond a discussion on perception and understanding. The discussion in this field has questioned the ability of people who are blind to appreciate any objects aesthetically, as it is assumed that without the channel of sight, such experience cannot furnish the mind. Beliefs about the nature of aesthetic understanding have also affected the way in which philosophers have thought about the moral state of the blind community, as many philosophers have also linked the appreciation of aesthetics with the ability to experience *decent* thoughts, believing that perception is linked to morality. Although these ideas now sound outmoded, they still remain at the foundation of much more recent educational, philosophical and latterly psychological discussion of blindness.

The existence of such paradoxes and ideas that do not reflect lived experiences highlight the problems with the power of academia. Those they influence construct what they consider to be reasonable ideas about what is possible for blind people to achieve, both in institutions and the greater society. Consequently, some of the problems that are associated with understanding blindness and the nature of perception are caused by those who promote themselves as the reformers of the blind community, and those who make decisions based on these beliefs.

The theories that have influenced this book's epistemological theory of blindness identify a broader method of defining this topic. More important, they describe the taxonomy of *attitudes* toward disabled people as an indicator of the treatment of disability as a social and cultural phenomenon, as well as a bodily one. For ease of identification, these theories can be roughly divided into two groups of literature: micro and macro studies of disability. Although this is an imperfect way of classifying these subjects, it eases an initial understanding of their modes of research. Micro studies of disability look at individuals' interactions with disability, and interactions of people who are able bodied with people who are disabled. These studies are termed micro as they are usually conducted through observation of small social situations, such as individual institutions that care for people who are disabled. Macro studies are observations of whole communities and cultures' understanding and treatment of disabled communities. They are largely conducted through an analysis of policies and literature, and are termed macro studies as they tend to make broad, generalized statements about their populations.

This chapter now continues by describing micro studies.

Micro Studies of Disability

Among the most significant and influential studies of interactions between people who are disabled are those written by the Canadian anthropologist Erving Goffman after the Second World War. In his seminal works on this subject, *Stigma* and *Asylums* (Goffman 1990, 1991), written from observations of small institutional communities of people with disabilities, he described the communal structure of a *local* social order. To Goffman, an understanding of how people feel they should act according to their social make-up was constructed through the witnessing of the social interactions between disabled and non-disabled individuals in confined settings. Interestingly, within this context Goffman sees physical disability in the same context as other individuals who are defined as living outside of what is considered to be normal society and prone to being deemed as outcast, such as criminals, or people who are mentally ill or have emotional problems.

In his analysis of behavior, objects and conversations Goffman regarded disabled individuals as symbols by which others defined their relative status. In this manner, by focusing on the extreme effects of social exclusion, disability was seen as social deviancy. Consequently he described the treatment of people with disabilities in the same category as other *social deviants,* such as criminals. This provided a notably broader social and cultural inspection of disability as a whole, rather than regarding it mainly in terms of the effects—be it related to appearance, physical indicators or otherwise—of a range of medical symptoms. He termed this process stigmatization.

Goffman not only observed the relationship between the wide communities and disabled communities, he also examined the reality of being a member of the institutions in which social deviants were housed. Such institutions, along with other forms of total institutions such as boarding schools and army bases, were defined as societies that were enclosed not only by physical precincts, but also by more abstract boundaries such as rules and customs. These institutions were places such as asylums for the blind, mentally ill or deaf (Goffman 1991). Surprisingly, in his observations Goffman found that despite stigmatization, many of the inmates of these institutions would be treated deferentially within its confines. On a one-to-one level, he observed that inmates and staff would often form a working consensus, and the boundaries between the stigmatized and the imposer of the stigma would greatly dissolve. Thus, in these institutional settings the stigma of being disabled was not quite so clearly defined.

In a similar manner, early social psychological essays on perceptual disability by the Austrian and Russian educational psychologists Berthold Lowenfeld and Lev Vygotsky focused on the use of social interaction in classes for children who were deaf or blind, and described the method by which education socially objectified disability. However, their stance on these observations differed from Goffman's. Although agreeing that social interactions played a large role in the behavior and learning beliefs of these children, they formulated a method of analysis that allowed for a duality of understanding the interaction between society and these individuals. For instance, instead of arguing that disability was deviancy, Lowenfeld and Vygotsky broke down the *context* of disability into two stages. The first was the notion that the medical symptoms leading to a disability are a condition caused by the symptoms of an illness, an undeniable medical phenomenon. The second stage was the notion that the school interacted with this disability using unintentional, arbitrary categories of symptoms. Thus, in this instance the brutality of definition was the problem, and the social outcome was merely a side-effect. It is thus the abstract definitions that disable those who do not fit the expected category of a normal person.

More recently, contemporary British disability theorists Colin Barnes and Geoff Mercer have categorized stage two of the process described by Lowenfeld and Vygotsky as a deficit model of thinking about disability. In this model, the stigmatization is swapped for a more oppressive form of a social quirk. This accepts a notion of social normality within an abstract notion of *perfection*, to the cost of those that are not normal. Although a more traditional source, however, Vygotsky's notion of this element of social classification has the greater clarity and purpose, as he describes the necessary dichotomy of the physical fact as well as the social classification of physical disability. This, he argues, causes socially disabling factors as a consequence rather than a direct cause of disability, as these more contemporary writers feel. Thus, he writes:

> It goes without saying that blindness and deafness are biological facts and not at all of a social nature, but the teacher has to deal not so much with the facts as with the social consequences of these facts. When we have a blind child as an object of education before us, we are compelled to deal not so much with the blindness itself, as with the conflicts which arise therefrom within the child when it enters life. . . . (Vygotsky 1994, pp. 19–20)

A later study on the culture of deafness by the American social historian Nora Groce (1988) in her book *Everyone Here Spoke Sign Language* and one by the British neurologist Oliver Sacks (2001) on visual impairment in his book *The Island of the Color Blind* provided evidence in support of Vygotsky's notion. They observed that interactions affect the way that disabled communities are normalized when they represent a significant element of the community, and only alienated when they are perceived as a small minority or their roles are marginalized.

Sacks observed a poor, inbred community on an isolated Micronesian atoll in the Pacific, in which total color blindness and photophobia were common. Groce on the other hand conducted a social history of inherited early and congenital deafness on Martha's Vineyard, New England, from the late seventeenth century until the early half of the twentieth century. Groce's fieldwork included collecting local documents and family trees, and interviewing the older members of the community. From this, she found that through this widespread use of sign language the abnormally large number of deaf people on the island normalized the definition of their biological condition's social standing to the point that deafness became barely a disability at all.

Sacks's later study followed Groce's original closely, although in his he was able to observe his community first hand as it still existed in the form it had for many generations. As a result, he was able to complete medical tests on those from the atoll on which this population was based. In his time on the Island, like Groce, Sacks observed that the condition of the significant minority of the Island had affected all of its inhabitants. Thus they represented examples of the esoteric exceptions to the mainstream social rules and, by extension, of the cultural nature of the construction, treatment and understanding of disability in communities where they represented a more significant minority. Above all, both writers found that the development of an autonomous community way of life based around the trait of sign language and a different use of light and color, became little problem for the community, both deaf and hearing, sighted and visually impaired, as they adapted to find a workable communication structure within their own communities.

So how do these theories explain a general theory of the epistemology of blindness? These scholars inform an understanding that, although based at some point in medical and biological assumptions of being less than fully able, the concept of blindness which is defined by what is presumed to be a common understanding is in fact defined according to societies, leading to a categorization of either disabled or non-disabled. Epistemologically, it could be said that such studies also inform our knowledge that definitions and diagnosis of blindness subjectify individuals, assigning features or traits according to a perception of shared characteristics. In such a way, micro studies look at the matter of the identification of characteristics in taxonomy, a mode of classification of blindness and disability. It is true, however, that subjectified societies do not exist naturally and there is not a natural understanding of their participants. It is this point that makes it imperative to not only comprehend the social condition of blindness, but also its process of definition.

The macro studies this chapter now turns to regard disability in such a manner. Their role in particular is to observe the cultural "concept" of blindness and disability, turning it into a commodity, a symbol of language or a representation. This is the most important function of the epistemological study described in the following chapters of this book. In doing so, they objectify a concept of blindness, reducing it to an allegory or metaphor of social existence, rather than an element of lived reality.

Macro Studies of Disability

Unlike the position of writers such as Vygotsky, Sacks and Goffman, the macro approach to disabilities is characterized most typically by the French historian and philosopher Michel Foucault (1989) and his analysis of mental illnesses in his book *Madness and Civilisation.* Although he does not focus on physical disability, it is important to understand his work in the general area of macro theories of disability as "object." Above all, it is important for having laid the groundwork for the field of the literary and historical discourse on disability.

Foucault described mental illness as a historical construction. This, he argued, was used as a function of power. Religion and governments have used the definition of madness to control deviancy—although like Goffman, Foucault took an extreme sociological position, their case studies show very different approaches of power and oppression versus a more arbitrary conceptualization of stigma. At the outset of his work, he found that the issues surrounding mental illness in different periods of history were unique, as they related to more socially constructed elements of disability. As a consequence, it changed its classification and diagnosis more frequently.

Furthermore, Foucault believed that the social relativism that surrounds mental illness has developed as a result of social changes throughout history. For instance, in the period after the Renaissance, people were confined as a result of poverty in urban environments. Even approaching the latter half of the twentieth century, he found that homosexuality was classified as a mental illness, and writers were considered mentally unstable for their outrageous academic beliefs. Thus, he suggested that what is assumed to be deviant, even evil, in one era can be regarded and acceptable, even glorified, in another. On this he concluded:

> Ruse the new triumph of madness: the world of thought to measure and justify madness thru psychology must justify itself before madness, since in its struggles and agonies it measures itself by the excess of works like those of Nietzsche, of Van Gogh, of Artaud. And nothing in itself, especially not what it can know of madness, assures the world that it is justified by such works of madness. (Foucault, 1989, p. 274)

Evidence supporting similar frameworks of objectified disability have been used to analyze blindness since the Enlightenment. Using a framework similar to the epistemological approach used by this book, the American academic William Paulson (1987) in his book *Enlightenment, Romanticism and the Blind in France* examined blindness as a concept of cultural narrative. He described the elevation of blindness in France around the time of the revolution, and its effect on the consensual attitude of blindness in the greater French society and their educational institutions of the era. Similarly, the American philosopher Norman Jay (1994) in *Downcast Eyes* depicted the same observation in twentieth century French literature, reaching similar conclusions.

Both writers found that French literature changed to accommodate the notion that sight was less important than generally received, as the sole channel of wisdom and morality. Both Paulson and Jay also felt that the denigration of sight in French culture was caused by a series of myths, which often distorted the effects of blindness. The source for this, they suggested, was epistemologically similar to the mechanism of ancient Greek literature—although this period took an opposite view of blindness as either a symbol of punishment or wisdom. Therefore, apparatuses exist in different cultures for the imposition of moral and ethical values on what is a physical condition. Paulson believed that this change of approach toward the theory of blindness occurred at the end of the eighteenth century. It was then, he continued, that the romantic culture of the Enlightenment in literature affected the development of an *open-minded* artistic attitude toward people who were blind. However, Paulson maintained that these attitudes then quickly became myths, and led to unrealistic attitudes about the ability of people who were blind, and glorified the effects of blindness in terms of innocence and moral virtue.

The Israeli art theorist and historian Moshe Barasch (2001) examined a similar concept in the mythological and symbolic representation of blindness in Western arts. He also focused on the formation of Christian Europe through the Bible and the literature of Antiquity. In general, Barasch identifies a change in the moral allegory of blindness during the Middle Ages and the Renaissance, finding that two prominent themes appeared during these periods. The first theme is the representation of blindness as a mythological punishment. The second is the image of the blind beggar being subsistent on those who are kind and charitable. Barasch's work also agreed with Paulson and Jay on the place of theology and the subjective nature of an understanding of blindness throughout history, and furthermore on the influence of ancient Greek literature on mythologies of blindness. In addition, though, he argued for a new form of discourse on our understanding of blindness as one linked to the cultural, theological development of a system of *narrowly interpreting* what is fundamentally a physical trait.

These writers' approach, however, differed distinctly from Foucault's insistence that the treatment of illness, particularly mental illness, was an active tenant of political power over those who suffered it. Instead, Barasch, Jay and Paulson suggested that this tenant was actually two strands of mythology, and a primitive theological notion of explaining the world. The first strand was blindness as a romanticized object of charitable deeds. The second was that people who were blind were evil and fiendish. The first strand is expressed more through activities such as giving assistance; the second is represented almost entirely through rhetoric and theological debate.

The role of theology in the understanding and role of blindness in Western culture was also the central theme of the British theologian and educationalist John Hull's (2001) interpretation of the Bible, through the focus of a man who is blind. In this study, he chose to highlight biblical myths of the creation of blindness, and argued that, similar to the findings of Paulson and Barasch in Greek literature, many had negative connotations. In particular, Hull saw blindness as a form of punishment of immoral acts, with a strong element of it related to sexual deviancy. He also often observed that linguistically the idiom used in biblical texts often favored sightedness to signify insight, indicating a cultural favoring of vision as the means of finding God and wisdom. Consequently Hull argued that in traditional theology the lack of light, and by extension a lack of sight, was often linked with doom and powerlessness in a great number of passages—in contrast to lesser instances in which darkness was often linked with calmness and being hallowed. This led Hull to question why the epistemology of darkness is so often linked with negative causes and consequences by many cultures, and in doing so questioned the dichotomy of images presented to the world through Christian literature.

Such a cornucopia of different approaches to the object or symbol of blindness described in these studies presents almost as a world apart from the lived realities of those who are blind, and the discussions of those who have observed blind and visually impaired people. In such a world, the notion of institutionalization of blindness is represented as a mechanism of definition, of mythology, of classification, and of morality. In these macro studies of blindness, its social concept is found to be one that emerges out of mythology, ancient beliefs and literature rather than scientific rationalization. In these works, blindness as a possession also becomes the custody of the institutional administrator and philanthropist, the Pope, the bishops, the theologians who write scripture, the administrators, the writers of literature, the painters and the philosophers. It ceases to be the property of those who live as blind people. Even so, like the findings of the micro studies discussed before, blindness becomes a social phenomenon, a label and a yolk that the barer had to carry.

What is perhaps most interesting in these findings, however, is that blindness is also seen as a cultural disability apart, defined by mainstream society as more dangerous than deafness or infirmity. People from a non- disabled world have chosen to isolate the lived reality of blindness as a special form of disability, and perhaps the most identifiable form of punishment or gift. The creation of a separate disabled community based on language differences, such as those who are born deaf or become deaf very early on, or even those that learn to read Braille, is well known, and as Groce's study in particular illustrated, is participated in readily by people from the disabled community. However, the historical and cultural identification of blindness, particularly as a removal of sight, is more telling as a starting point of an epistemological theory of blindness. It is at this point, therefore, that this book begins its study of the creation of an epistemology of blindness from the beginning of the Enlightenment to nationwide integration in schools; the period during which our understanding of blindness changed, and developed into the symbol and object of blindness that we know today.

Before beginning this task though, this book reviews its following chapters as a guide for the reader.

THE FOLLOWING CHAPTERS OF THIS BOOK

This book is broken into three parts. Definitions and studies of blindness and touch are presented in the first part, which is split into two chapters. Although this book focuses on English attitudes to blindness, this study broadens to those from other parts of Europe and North America. Research in this field cannot be seen in isolation, as the findings are so intertwined. Although we show there are cultural quirks to these studies, it is impossible to see the development of the study in England without reference to these other interpretations.

More specifically, these chapters discuss the epistemological history of the earliest studies in the Enlightenment up until those in the early twentieth century. They also feature a study of the modern definitions and the early history of the philosophy and medical psychology of blindness, and their motivations. This part also reviews the study of blindness and touch in what we call the modern era, from the middle of the twentieth century onward. It includes psychological, philosophical and educational studies, and studies of language and communication. The aim of this part is to provide a grounding in academic attitudes to blindness, and to show how it has affected the accepted wisdom about blindness.

In the second part of this book we describe the development of institutionalized education in England from the late eighteenth century until the 1981 Education Act. Again, although these focus on England, we also make mention of other educational philosophies, in order to place the English experience in its greater context. This description is split into three chapters. These examine the history of touch in vocational and aesthetic education in English schools for the blind, and the legal process that led to integration in 1981.

In particular, this part presents an analysis of the social and cultural development of the hand crafts, music and the touch curriculum for the blind throughout its one hundred and ninety-year history, from the first asylum in 1791 to the publication of the Warnock Report and its consequences. In this part we also present an examination of the moral use of aesthetic education of the blind. The aim of this chapter is to show the motivations behind English education of the blind, and particularly how external social and cultural factors have influenced the creation, management and curriculum of these institutions.

Finally, in the third part of this book we present our conclusion in a single chapter. In this part, we also present a broad set of recommendations for the future.

We now continue with the first chapter in part one of our book.

PART I

CHAPTER 2

DEFINITIONS AND THE EARLY STUDIES OF BLINDNESS AND TOUCH

INTRODUCTION

This chapter describes how blindness is currently defined, and a contemporary history of the philosophy and psychology of blindness and touch. This latter study can be divided into two models: cognitive and perceptual studies. Literature in these fields reveals that these have a common root in the late seventeenth and eighteenth centuries philosophical understanding of blindness, during the early years of the Enlightenment. In many cases, these studies stem from an early question by William Molyneux. Furthermore, in common with many other academic disciplines, both models of research focus on the medical classification of blindness, and many of these forms of study revert to what are now regarded as legal definitions of blindness.

What follows in this chapter is a study of the epistemological evolution of academic studies, which provide archaeology of attitudes toward blindness. This is split into two sections. The first examines the definitions of blindness and disability, handicap and impairment, on which our modern notions of blindness are based. The second examines the traditional strand of research in this field, and discusses research into the nature of perception.

We now continue with a discussion of the definitions.

God, Money, and Politics, pages 17–35

DEFINITIONS OF BLINDNESS

The precise delineation between people who are blind, those who are sighted and those that are visually impaired is confused by different regional diagnoses, although in recent decades this problem has been lessened by more accurate scientific definitions. This previous lack of cohesion makes it difficult to make a totally accurate comparison of studies, and a consistent study of blindness and art. However, each definition appears to describe visual acuity, the ability to distinguish between two visual points at which sight is usable when reading; full sight is defined as the ability to read a whole standard chart, and any loss thereafter is expressed as a fraction of this perfect vision (Feinberg, 2003).

After the diagnosis of lost visual acuity—the blindness—medical studies then sub-classify blindness according to either its effect on the body, which is its change of visual perception, or the causes of the disability. These definitions are then further subdivided into the stage at which a person is blinded, and also determines how blind people understand their remaining senses (Feinberg, 2003). Each of these approaches is now discussed under two further subheadings: perceptual definitions of blindness, and developmental definitions of blindness.

Perceptual Definitions of Blindness

The World Health Authority's (WHA) modern definition of blindness is $3/60$ visual acuity and below (Whitcher et al., 2001). People who have usable sight but nonetheless have lost a significant amount of visual acuity are defined as visually impaired. Yet this definition is ambiguous and remains nationally subjective, as so many further forms of visual impairment exist that are not just based on visual acuity; these include photophobia and color blindness. Furthermore, problems in the methods of calculation also exist to further complicate this issue. The British Isles, for instance, uses the WHA's definition, and the government classifies blindness as 5% sight, or $1/20$ visual acuity. In the United States, this definition is adjusted to 10% sight, or $2/20$ visual acuity in the eye with the greatest amount of vision. Although these definitions could be seen as amounting to a similar amount of vision, in reality the structure of blindness is vast, as 10% vision in one eye can allow usable vision for reading, whereas an overall visual acuity of 5% can diminish this aptitude radically.

The purpose of these definitions, according to medical rhetoric, is to treat the diseases causing blindness or to alleviate the extreme elements of sight loss. According to the British ophthalmologists Richard Coakes and Peter Holmes Sellors (1992), these treatments either can be used as a guide to the cure for the blindness itself, such as through the removal of cataracts, or to alleviate the worst effects of blindness. However, as the British neurologist Oliver Sacks (2001) found in the case of traits such as achromatism,[1] this definition can become further complicated as visual acuity can increase through lack of normal light, and yet the testing of this visual acuity must take place in what are considered to be normal lighting conditions. Therefore, the whole of the person is judged according to a rigid scientific/medical test at a particular point in time, under certain conditions.

In addition to its definition, blindness is also scientifically explained as a disability, handicap or impairment, depending on the strength and form of impediment to the functioning it is felt to cause. Slightly confusingly, these definitions are produced by the World Health Organisation (WHO), which, unlike the scale used for classifying legal blindness, appear to describe all three qualitatively. The WHO's (1980) official definitions find that disability is related to a belief in the existence of social norms, and their classification of impairment is more related to an intrinsic "disfigurement" of the body or mind, whereas handicap is more related to social attitudes linked to the outcomes of disability or impairment. This is set out in a series of documentation for guidance:

> **Impairment**: Any loss or abnormality of psychological, physiological or anatomical structure of function.
> **Disability**: Any restriction or lack (resulting from an impairment) of ability to perform an activity in the manner or within the range considered normal for a human being.
> **Handicap**: A disadvantage for a given individual, resulting from an impairment or disability, that limits or prevents the fulfilment of a role (depending on age, sex, social and cultural factors) (WHO, 1980, p. 29).

These are controversial beliefs. The British disability theorists Colin Barnes and Geoff Mercer (2003) argue that the WHO's definitions of disability, handicap and impairment can become disabling issues themselves. Such definitions become imposed on those they are aimed to help, and lend an official, institutional aim to their diagnosis that seemingly defines a cultural *truth*. Furthermore, Barnes and Mercer claim that these definitions are based on a rigid belief in the ability of an individual to complete tasks as well as fulfil social, cultural and economic roles. These, they argue, are based on the imposed social and cultural assumptions of a powerful administrative body, which may change according to region. Therefore, such definitions themselves can affect the attitudes of a society when encoded in law, as they classify what are expected to be normal functions within individual cultures.

Political and medical definitions can also have the effect of imposing what the Canadian anthropologist Erving Goffman referred to as a stigmatizing effect on the individual, affecting the blind person's psyche, and further delineating them from what it feels like to be *normal* in society. This section now continues by examining further psychological classifications based on the timing of blindness and its effects on the mind.

Developmental Definitions of Blindness

Developmental psychologists delineate between the ages at which people become blind, as this affects the manner in which people who are blind cognitively interpret their perceptions. In the early development of this study in the 1950s and 1960s, the Austrian psychologist and educator Berthold Lowenfeld (1981) defined four general categories of behavioral delineation, not just based on the levels of sight but also on the age at which blindness occurs. These state that personal experiences of perceptions in early life are of particular importance to cognitive and emotional development. The levels are:

1. Total blindness, congenital or acquired before five years of age.
2. Total blindness, acquired after five years of age.
3. Partial blindness, congenital.
4. Partial blindness, acquired [after 5 years of age] (Lowenfeld, 1981, p. 67).

Furthermore, Berthold Lowenfeld argued that children who were congenitally blind should be seen as an entirely different category to each of the others. These children, he wrote, are the only category of person that has never experienced light perception, and therefore have a completely different frame of reference in language and intellectual development. More recently, the ability to define broad physiological and psychological categories, such as Lowenfeld's, has been questioned in contemporary studies.

In particular, my own findings (Hayhoe, 2004, 2005a,b, 2006) and those of the American psychologist David Warren (1994) in the study of educational psychology have argued for an understanding of mental influences beyond those of a simple medical and chronological categorization. Both of us have claimed that there are so many different influences on a blind child's development during their early years, including their social and cultural background as well as their intellectual influences, that it appeared to be impossible to classify such students with any meaningful accuracy. Thus, perhaps what is most needed is a model that approaches the definition of people who are blind not as elements of a category, but as individuals with different physical, social and emotional *traits.*

Nevertheless, despite models which take account of non-clinical factors, mainstream research into blindness, art and aesthetics continue to use their own categories of blindness based on loose groups of what seems to be a perceptual commonality. These definitions originate from a need to understand blindness not just as a physical phenomenon but also as the result of a disease. As previous studies have discovered, blindness is unique in this way and lends more to its challenge to the supremacy of sight by the political establishment, as a symbol of radicalism. In this context, it would later become a challenge to a maintained order of states, and a means by which the whole notion of intellect and thought could be challenged.

Contained in this notion was a tacit suppression of aesthetic understanding and an acceptance of morality through iconography and vision was to equally become the tenant of the social conservative. Thus, as the notion of the understanding of blindness was to the radical, the control of this symbol or object of blindness became as important to those who wished to maintain the social order. The review of the roots of this philosophy and the research that followed it is now discussed below.

STUDIES OF THE NATURE OF TOUCH AND BLINDNESS: THE FOUNDATION OF THE METHOD

In the twentieth century, the British psychologist and philosopher Richard Gregory (1987) and the American linguist and philosopher William Paulson (1987) traced the beginning of the contemporary study of blindness to the earliest works of the English and Irish enlightenments. Similar contentions were made in the nineteenth century by the American educator Samuel Howe, and also in the eighteenth century by the Irish, Scottish and French social theorists George Berkeley (1899), Demodocus (anon., 1774), Valentin Hauy (1889) and Denis Diderot (2001). Each of these writers concur that this revolution of thinking began with the question featured in our introduction, to the English legal and moral philosopher John Locke by the Irish natural philosopher William Molyneux.

The question was later reproduced in edited form under the heading "Of Perception," in Locke's seminal work *An Essay Concerning Human Understanding*, in its second edition (2001). This he published in late 1690 on his return to England after a period of political exile—according to the American philosopher John Yolton (2001), the first edition was published in French in 1688, and thus the following correspondence was written between the publications of the first two editions as an ongoing process of philosophical exploration. At the time, this form of study remained untested by natural philosophers, as surgical procedures did not exist to rectify this condition. In its range, however, this question did not simply address the narrow study of the relationship between touch and sight. It was fundamental to understanding the nature of the perception and comprehension of what was later to be described as the Humanist understanding of existence.

The focus of this study was also not simply inspired by altruistic, medical or scientific motivations in order to frame this far-reaching approach to thought. Neither was the use of blindness solely a means of understanding cognitive mechanisms, as the British psychologist Richard Gregory (1974) later maintained. Its origin was part of a more complex scheme that questioned what was known in the human mind, that questioned the role of the orthodox *fundamentalism* of the Anglican Church in England, and perhaps more important, that undermined the theological *fundamentalism* of the Catholic Church.

To understand these motivations in greater depth, it is important to recognize that the hypothetical discussion of the *blind man* did not exist in a philosophical vacuum within the essay, or indeed in the studies that resulted in this work. In addition, it is essential to realize that the illustration of blindness in the original letter to Locke was only part of his original analysis of the experience of blindness in the essay.

We continue by discussing these elements.

Locke's Use of Blindness in Understanding Philosophy

In an earlier chapter of the essay, Locke used an example of a blind person whom he had interviewed in order to understand the difference between innate and learned principles of the mind. This discussion also emanated from a general debate that was being held between radical thinkers, some identified and others whose names were kept secret, the most infamous of which was Isaac Newton. It was this discussion that led to his original drafting of the essay.

In the context of human understanding in this period, Locke argued that traditional human knowledge of spirituality, as described by the orthodox forms of Christianity, acknowledged an inward understanding of divinity, one that was born in an individual, and one that was enhanced by a vision of the world around. As the Israeli historian and British theologian Moshe Barasch and John Hull, respectively, identified, such feelings were not uncommon in early Jewish, Christian and Greek spirituality, or in the academic works of the Middle Ages and the Renaissance.

Rather than questioning this notion of blindness as a denial of divinity, and thus as a punishment of sin, Locke interrogated the core understanding of what could be perceived as Godly by examining a fundamental tenant of an orthodox Catholic belief: that morality was intrinsic. Under the heading Further Considerations Concerning Innate Principles Both Speculative and Practical, he asserted that what was *felt* was actually *learnt,* and then *perceived* through all modes of understanding and sensation. More important, in this chapter he identified memory as the foremost mode of understanding, rather than a more inward notion of deliberation. This was previously held to be the key to human understanding. As he stated in his text:

> [A] blind man I once talked with, who lost his sight by the smallpox when he was a small child [had] no more notion of colours than one born blind. I ask whether anyone can say this man had any ideas of colours in his mind, any more than one born blind? And I think nobody will say that either of them had in his mind any idea of colours at all. . . .
>
> [The] truth is, ideas and notions are no more born in us than arts and sciences, though some of them indeed offer themselves to our faculties more readily than others and are therefore more generally received, though that too be according as the organs of our bodies and powers of our minds happen to be employed: God having fitted men with faculties and means to discover, receive, and retain truths, accordingly as they are employed. (Locke, 2001, pp. 41–42)

This was a highly controversial claim to make at the time. It was also indicative of Locke's own theological, and ultimately political journey. He was the son of a Puritan soldier from Somerset, a particularly rebellious county at the time, who had fought as a member of Oliver Cromwell's army during the civil war. These religious and political family connections helped Locke to gain a place at Westminster School in the first instance, and then later secure a scholarship to Oxford. In his early years at Oxford University, Locke also contributed to a book of poems praising Cromwell.

However, as the American historian William Spellman (1997) found, when Cromwell's death led to the cessation of the Protectorship and the restitution of the Church of England, Locke, now a lecturer at Christchurch College, Oxford, believed in a politically unified theology and a united front for the Anglican Church. More important, he was radically opposed to dissent of any kind at this time. This was an era of great confusion and political upheaval in the country, and in this malaise the Anglo Catholic Church held a pivotal role in maintaining cultural and social stability.

After the restoration of the monarchy to England, Charles II began taking the country's governance back toward conciliation with Roman Catholics and more orthodox Anglo Catholics. During the latter years of this era, influential academic leaders were compelled to reflect this doctrine; for example, Spellman finds that all Oxford and Cambridge colleges were forced to have Catholic masters. As a counterpoint to this position, many powerful figures in government and academia questioned this shift in position, and formulated counter theologies and political ideologies. It would appear, therefore, that this version of Gnosticism not only held spiritual meaning to its creators, it also acted as a political symbol of radicalism and insurrection toward the monarch, whose popularity was tenuous.

Although Locke had initially sided with this orthodox theology, after meeting many of the well known religious and political radicals at the time of the restoration, he shifted his stance to sympathize with his family background of political radicalism. Although staying loyal to the broader cause of the Anglican Church, he formed a circle of friends with radical intentions, who were to become members of his philosophical circle, and later had influence on the drafting of his Essay. Because Locke associated with such political and religious activists, in particular the influential Earl of Shaftsbury who was to eventually plot against the monarchy, an order was made for his arrest and he was forced to flee England for Holland in 1683. A year later, Charles II publicly expelled him from Christchurch College, Oxford.

Rather than dampening his influence, Locke's arrest warrant only served to heighten his political and academic influence, and in the remaining years of the seventeenth century, Locke became an academic focal point for what was the theological and political counter culture in England. It was also during this period that he published his first edition of the Essay. Pertinently, in the years leading up to its second edition, Locke engaged in a secret theological dialogue with Isaac Newton, who also challenged the notion of the primacy of sight, and more particularly the place of vision in many passages of the Bible. This association also appeared to be stronger than the intellectual friendship between Locke and Molyneux, and represented a motivation that was to prove a more damaging threat to Anglo and Roman Catholic theology.

This more covert line of discussion is now discussed below.

Isaac Newton's Study of God and Light

Isaac Newton's serious study of theology happened in the middle years of his academic career. As a condition of election to his fellowship of Trinity College, he had to be ordained as an Anglican priest. In doing so, the American historians of science I. B. Cohen and Richard Westfall (Newton, 1995) found that in the process of studying for this ordination Newton applied the same rigorous methods of criticism and attention to detail that he did to his more famous studies of mathematics and physics. During his tutelage and copious reading of the history of Christianity and scripture, Newton found what he felt to be many discrepancies in the early interpretations of the Bible, which, he argued, were concocted as a historical plot by Roman supporting, Latin speaking scholars from the time of Constantine in the fourth century to the then present. In particular, Newton argued that several of the *rudimentary* tenants of orthodox Anglicanism and Catholicism, such as the Trinity of God the Father, Son and Holy Spirit, were in fact a corruption of the original scriptures.

Newton continued that the notion of the Trinity was introduced at early biblical councils, by those interpreting scripture into Latin for the purposes of winning and retaining power for the Papacy and Rome. These scriptures, he contended, were changed from an original translation of scripture by Gnostics such as Arius, an Alexandrian Deacon of Libyan decent from the fourth century, who was one of many influential scholars of Greek translations—indeed Newton's faith became influenced by this Gnostic. The original confession of faith by Arius provided a more secular translation of the role of Jesus, and even doubted that he possessed the Holy Spirit. This form of belief had subsequently been declared a heresy by Roman Catholic authorities, his books were burnt, and any sympathy toward his ideas was also declared a heresy and thus punishable by excommunication. For Newton to hold such beliefs was dangerous, particularly as he studied the Bible in a period of orthodoxy in English religion.

After making his acquaintance in the late 1680s, Newton's development of this study led him to send two discourses of these and similar findings to Locke for anonymous publication under the title *Letter to a Friend* (Newton, 1959–1977). These first works were subsequently published in France by a friend of Locke and were followed by a third letter, which was unpublished until its presentation in a later collection of letters in the twentieth century, as it was felt that the first two had sufficiently endangered Newton's position at Cambridge University.

The timing of the later publication of Newton's letters was again significant. They came shortly after Locke had returned from exile in Continental Europe. The new Monarchs in this period, William and Mary, were attempting to reassert a more distinctively non-Catholic form of the Protestant faith in England. The suppression of orthodox Catholicism now held political as well as theological sway in English government. Radical scholars were discovering an England where, although still considered alternative, nonconformist Christian views were accepted and benefitted from a less threatening environment. It is not beyond reasonable consideration, therefore, that those who had been previously hunted or marginalized by the establishment, including Locke and Newton, would have been particularly resentful of those elements of scripture.

As a result of what he saw as this Catholic corruption of theology, Newton in particular turned his resentment of the Catholics to their symbols of power. For example, Newton particularly disliked idolatry and the use of gratuitous visual ornamentation as objects of worship. Later, as Master of Trinity, he would have the stained glass windows in the college replaced with plain glass. Furthermore, as evidenced by further private correspondence with Locke in the early 1690s,[2] Newton also scientifically doubted the feasibility of visions. This particular discussion with Locke focused on the biblical passage "The Ancient of Days" (Daniel, Chapter 7), a further tenant of the belief of orthodox spirituality, in which the angelic vision is observed from a bright light in the sky.[3] As Newton describes in one such letter, after attempting a scientific observation in which he exposed his eyes to strong sunlight for a protracted period, and feeling a state of emotional turmoil, he could himself reproduce a similar apparition *in his own mind.* This, he concluded, brought into question the metaphysical nature of perception, and its subsequent connection with God.

More interesting, he continued by describing the feelings he had when finishing this experiment and retreating from the sun into a darkened room. It was here that he again conjured the apparition in his mind. This, he felt, suggested that the visions described in the passage could have been imagined as a psychological "phantom," as he himself had pictured it. This in itself would appear to conform to Locke's understanding that all notions of color were in the mind, and the physical use of the senses was simply a mental occurrence rather than, as orthodox texts had it, a consequence of an inherent acceptance of God.

Newton's discussion of this observation was also timely as it was in the era in which he and Locke, using different modes of thought, were beginning to seriously examine the tenants of the theory of color by the English chemist Boyle—who hypothesized that colors could be reproduced and altered through chemical composition. This debate between Newton and Locke led to the idea that color was therefore produced physically and not metaphysically, its changes tested by empirical methods, and that the memory of color could be conjured from consciousness rather than being an inherent notion of vision. For Newton, this belief was mirrored in his theory of optics, in which he proposed that light was composed of particles that could be refracted and controlled.

It was thus that light and vision not only became symbols of scientific speculation more than any other form of perceptual understanding; in the mind of Newton they also became political and theological symbols of orthodoxy and fundamentally held beliefs. This discourse would change little in following centuries of study, during which research into these phenomena, and policies based on their understanding, would be tested. The political and academic necessity of these studies made the understanding of blindness a symbol of academic intrigue over all other physical disabilities.

It is these empirical studies that this chapter now addresses.

Early Empirical Studies of Locke's Ideas

According to Paulson (1987), Locke's discussion of vision and blind people contributed heavily to the growth of interest in the scientific study of blindness as a human state, and light as a psychological phenomenon. This led to a more widespread discussion of its form in many European cultures. Such a debate also influenced the study of art in the eighteenth century, and as the British art historian David Bindman (1981) argues, under the English painter and cartoonist William Hogarth in particular. It was Hogarth, the first president of the Royal Academy and one who also criticized the Catholic faith through his art works and essays, whose reading of Locke inspired a new form of aesthetics. This was to eventually lead to the continental European fine art movement of Modernism. In terms of its symbolic value, this new consideration of light and vision was also a strong statement that sight would have to be learnt and that, like an understanding of color, it was not inherent.

In terms of its effect on medical research, initially the most influential field of study of this topic, Locke's reproduction of Molyneux's question, was empirically tested less than two decades after its original publication. In 1709 the Irish philosopher and physician George Berkeley (1899) was able to study a young man who had just had congenital cataracts removed, a surgical procedure that had only recently been invented. During observations of this man, Berkeley discovered that, although he did not recognize items initially by sight, a short period after he touched objects and simultaneously saw them he could then recognize the objects by sight alone. As a result, he offered an initial conclusion to this question, feeling that people born blind would awaken without an understanding of their conditions and environment only at first. These would then have to be *learnt* through a mental process and not derived through inherent means.

> [A] man born blind, being made to see, would at first have no idea of distance by sight; the sun and stars, the remotest objects as well as the nearer, would all seem to be in his eye, or rather in his mind. . . . (Berkeley, 1899, p. 187)

Although the philosophical conjecture of touch and total blindness continued for more than two centuries, it was not until the twentieth century that a structured, empirically testable, scientific and above all psychological theory of nonvisual touch aesthetics developed. Although there is no specific structured discourse of the development of this study, it seems that this theory converged from studies in different cultures of understanding. In the United States, a psychological theory of touch vocabulary was proposed by the American psychologists David Katz and James Gibson, which was fundamentally distinct from previous medical studies (Schiff & Foulke,

1982); although again such studies took a great deal from theories relating to Phenomenology, the descendent of Humanism. In the same era in Europe, an empirical study was developed by the German psychologist Max von Senden, which borrowed to a large extent from the observations made in Berkeley's earlier analysis and methods (Gregory, 1987).

There were also lesser known documented educational experiments which featured case studies of students who were blind, using touch to understand artifacts from early in the twentieth century. These stood apart from their psychological counterparts not only in their method, but also in their belief in the ability of people who were blind to gain some form of aesthetic understanding from objects. The first of these studies was by the British curator Edward Charlton-Deas (1913) in Sunderland. It appears, however, that its empirical nature can be seen as relative to those of other studies in the twentieth century, which were conducted in laboratory settings. This more subjective observational study involved descriptive interpretations of students' work from local schools for the blind, after they had been led around museums and galleries and then asked to re-create the objects they had experienced.

In his conclusion to a comparison of similar later exhibitions, the British curator Bartlett (1955) found that students showed signs of not simply an aesthetic understanding of gallery artifacts, but also an appreciation of these objects and an ability to understand their material and spatial artistic properties. His literature search also concluded that much could be gained from museum courses based on similar methods in the future. However, he also found a more negative theme in this literature that appeared to suggest an institutional resistance to introducing such courses into museum and gallery spaces on a regular basis. This led him to the conclusion that institutional authorities did not like the idea of blind people engaging closely with aesthetic objects.

Regrettably, these studies did not lead to a greater corpus of knowledge about the appreciation of artifacts by people who were blind. Nor does their documentation appear to provide a clear, replicable methodology, or a delineation between or definition of students who were partially sighted and those who were blind. Such studies still worked on the premise that blindness was the removal of all light, and an understanding of light. Furthermore, according to the American psychologists William Schiff and Emerson Foulke (1982), it was also not until Katz's later research on the physiological character of touch emerged in the first decades of the twentieth century that it was acknowledged touch could be investigated as a controllable psychological phenomenon, providing information with characteristics similar to those of its peer senses. Indeed, Katz's studies were to later inspire discussion by Berthold Lowenfeld (1981), which would later still encourage a train of developmental psychology concerned with the welfare of children in schools for the blind.

However, in this early era of psychological study it was von Senden's analysis of previous case studies of blindness and the understanding of objects, published originally in German in 1932, that produced the more extensive body of replicable theory (Revesz, 1950). This literature, like its orthodox predecessors, continued to distrust an ability to perceive beauty, and doubted a learnt ability to adapt to a nonvisual creative understanding of the world. It was not until the 1950s that this thought process appeared to change. This would come with the advent of a work initially of criticism and them empiricism, by the psychologist Revesz.

In his book *Psychology and the Art of the Blind*, the Hungarian psychologist Geza Revesz (1950) found that, unlike Berkeley's earlier study, von Senden compared *published* records of subjects who were blind and had subsequently been cured through surgery, rather than structured observations of patients. Von Senden's findings, although vague, also appeared to contradict Berkeley's earlier, more reliable discovery. As he concluded, the sense of touch created little or no part in their conception of objects or space, and so Revesz heavily criticized this study as lacking both empirical rigor and an understanding of the different developmental categories of blindness. The excerpt below from this book illustrates this:

> Senden (sic.) has further paid no attention to the time that has elapsed between the operation and the testing, nor to the presence or absence of any psychological data about the condition of the patient before the operation. (Revesz, 1950, p. 15)

Later writers thus found that Revesz's own empirical study, which carefully documented an empirical methodology for the study of touch, blindness and aesthetics, later became more important than von Senden's (Gregory, 1987). In it, Revesz combined an evaluation of von Senden's case studies with observations under controlled laboratory conditions with people who were early blind, blindfolded and sighted, who all touched and reproduced artefacts (Revesz, 1950). Revesz's analysis included a number of theoretical areas: spatial perception, haptic (active touch) perception of objects, and the nature of aesthetics and creativity. Furthermore, in a direct comparison with von Senden's study, Revesz also later reviewed a number of historical case studies of artists who were blind or partially sighted. These case studies described artists practicing over a time span of one hundred years in Continental Europe, although all but one of these was late blind—he made this choice of subjects as these were the only reliable case studies available.

Most of Revesz's data is presented as verbal observations, artifacts that were produced during the experiments, photographs of the actions of the creative process, quantitative measurements of making times and the ratios of the objects made to their copies. His analysis of this datum suggested that many of the negative assumptions of von Senden's should be rejected, based on two main arguments. The first was that there were two types of artistic viewers: the haptic type and the visual type—those that enjoyed touching objects and those that relied mainly on sight; although these two types of viewers did not depend on their level of sight. Revesz's second finding was that the viewers who were blind had a form of spatial perception and could understand the objects they were given; in particular, they identified their features and differentiated between elements of different pieces.

Despite his disagreement with von Senden's analysis of haptic spatial perception, however, Revesz did agree with von Senden on a major point. He rejected any possible notion of an aesthetic appreciation of beauty, or the creation of images of *truly* aesthetic beauty, without any understanding of visual memories. In particular, Revesz concluded that the only sources of beauty that could be transmitted from artists to viewers were color and allegorical symbolism. The excerpt below from his essay illustrates this point:

> [From] what sources could a blind person, who has never seen the world with all its wealth of forms and color, derive those manifold experiences? . . . [No] one born blind is able to become aware of the diversity of nature and to apprehend all the rich and various appearances of objects. (Revesz, 1950, pp. 316–317)

CONCLUSION

Our evidence shows that the way we define, classify and develop our understanding of blindness is strongly governed by social and cultural factors. As importantly, this cultural understanding of blindness in England has evolved over a period of centuries, and began under the influence of external religious and political factors during the period of the Enlightenment. These factors are now discussed below under separate subheadings.

The Subjectivity of Definitions of Blindness

There has never been a single, international understanding of blindness, or indeed of who *blind people* are. And it is arguable that as long as *deep seated* cultural differences exist there can never be one. These differences are largely motivated by different social as well as academic paradigms, and the need of quasi political institutions to reduce the many characteristics of blindness to a single, distilled equation of what it *means* to be blind. Thus their use of reductionist terminology and supposed empirical testing conditions during diagnosis produce a vastly oversimplified understanding of the varying traits or forms of blindness, such as color blindness, tunnel vision and cataracts.

It is also apparent that the medical definitions we examined in this chapter are over-reliant on an image of blindness as a set of symptoms that can be cured, or at least treated within a *therapeutic* framework. Thus blindness is seen as akin to illness, leading to a form of whole body deficit to the person who is blind, as well as a sense of physical pain and discomfort. This too often contributes to an idea that blind people are objects of treatment and in extreme cases charity, rather than those whose traits can be circumvented through practical measures when *specific* symptoms are sensibly identified.

Unfortunately, these medical analyses are too often legitimized by international authorities, such as the WHA, who have traditionally attempted to categorize blindness through a process of negotiation and a reduction of a further range of international medical definitions rather than through a holistic, interdisciplinary discussion about the reality of what it is to be blind, although this approach is also in danger of pigeonholing and reducing the greater understanding of what functional vision in a broader cultural context is, if it seeks to oversimplify these wider notions. In many Western cultures in particular, these definitions are often manipulated by the ability to work and study, and divide those that receive state assistance from those who are deemed *fit* to work. Consequently, they attempt to impose a more general social classification of deficit, in common with many others who have physical and psychological impairments.

The dangers of restricted classifications have been compounded more recently with a trend toward the further reduced definitions of what it is to be disabled. These have been encouraged by international, quasi-governmental organizations such as the WHO, who have attempted to include social and psychological factors, but have fared little better in their classification of blindness within a three-tiered framework of Disability, Impairment and Handicap. Indeed, these attempts have turned out to be even simpler than the medical classifications they are trying to improve, as they are trapped in a quagmire of competing academic and governmental paradigms. In effect this process unnecessarily condenses blindness to a wholly unsuitable piece of an ill-fitting jigsaw, which is solely factionalized according to a cultural need to define labor and educational attainment.

This conclusion now continues with a discussion of the affects that religious and cultural factors had on the study of blindness in England.

The Affects of Religion on Our Cultural Understanding of Blindness

The concentration of modern, empirical studies of blindness on comparative perception—that is nonvisual perception versus visual perception—is not wholly based on practical scientific logic. Neither is the apparent concentration on touch. This deduction is the product of an era that became obsessed with challenging traditional conceptions of religion and its effects on political power, in which academia took a leading role in providing ammunition for a theological war. For Locke this link was concretely demonstrated with his expulsion from Christchurch College, and his exile in continental Europe.

Our discussion has shown that, unlike the presumptions by later generations of philosophers and psychologists, this question was not asked in isolation and was not the only instance of a consideration of blindness by Locke or his circle of philosophers. The experience of blindness is featured throughout his essay, discussed and examined many times more than any other form of perceptual impairment. Although we cannot be sure why blind people were chosen above all others, there is certainly circumstantial evidence that this was part of a greater debate, involving far broader theologically and scientifically more celebrated subjects: optics, metaphysical visions and idolatry.

Although Isaac Newton could not be named as one of Locke's inner circle of philosophers, at the time of the publication of the second edition of his Essay, it is without doubt that his opinion was admired by Locke. If it had not been, Locke would not have subsequently risked his reputation by urging the publication of his Letters to a Friend. He would also not have later entered into a debate with Newton on the biblical passage from the Book of Daniel, The Ancient of Days, when he was still in a precarious political and academic position.

Thus it could be said that the motivation for his poignant discussion on blindness was in part to show the lack of an *inherited* ability to understand visions and icons, as orthodox Catholics believed was the case. Thus, its essence also challenged the primacy of sight as the foremost channel of understanding God, and all Godly things. For Newton, an Arian who despised the idea of the Trinity as a corruption of true Gnosticism, and idolatry and all of its visual references as profane, his discussion with Locke about Boyle's book of colors presented a route to proving a theological as well as a scientific point. Thus, Boyle's book would become a major influence on one of his most famous works, *Optics*, published in the early eighteenth century.

Thus it was that blindness was a safe idea for these thinkers to hide behind. At this time of political and religious turmoil, when idle thoughts could lead to imprisonment and the loss of a profession, philosophers could use the symbol of a blind man as an academic metaphor. In a similar manner, it was also his challenge of innate ideas, in the same manner as his secret soul mate Newton, which demonstrated a cynical, Gnostic core to Locke's thinking.

Under the final subheading in this section, we discuss how touch became the symbol of the study of blindness.

The Use of Touch in the Study of Blindness

Following the period of the early Enlightenment, the currency of academia changed. The use of the political and theological symbolism of the previous century became legitimized in the minds of those that followed. It was fortuitous that Berkeley was active in research at the time of a surgical technique that cured some forms of congenital cataracts, as he was one of the early pioneers of what would later become the empirical psychological method, a mode of testing that would legitimize this science to its loftier place in academia, and raise it above a simple branch of the philosophical sciences.

Thus, the use of blindness and touch as a symbol of pioneering academic thought also elevated blind people to objects of interest above all other categories of disability. It made the plight of the blind an infamous object of fascination, like the Galapagos Islands to Evolutionists, for centuries to come. Unfortunately, this brought prejudice as well as wisdom. Although the early observations of Berkeley shed light on learned perceptions, they also cemented a distance between ensuing writers: psychologists von Senden and Revesz restricted much of their subsequent studies to data from medical and laboratory studies, doubting the abilities of blind people, and their work was discussed in later studies; the curator Charlton Deas worked with blind people and saw a greater ability first hand, although his observations were ignored by scientific studies that later shaped policies on education and social welfare for the blind. Because these paradigms could not mix, orthodox empiricism was favored above all observations, and academics could not build an early accurate picture of what it was like to be blind.

It was from these early twentieth century studies that two further rhetorical themes emerged as a topic of study: the observation of touch in education, and the biographical accounts of personal experience—including the analysis of classroom observations and informal experiments in classrooms. The pedagogists who conducted these studies, unlike their laboratory-based peers, had worked with students who were blind and had become increasingly persuaded by their artistic ability. As we have shown, this body of work was little recognized, as it was the empiricists who were to create the corpus of theory that could be methodologically reproduced and thus discussed more *accurately*.

In the next chapter we discuss these studies, produced in the latter half of the twentieth century, which referred to these earlier models while effectively ignoring what we now realize are more accurate findings.

NOTES

1. Achromatism is the inability to see any colors other than black and white, and leads to photophobia.
2. Correspondence of John Locke, the John Locke Collection, Bodleian Library, Oxford University.
3. The observation you make in Mr. Boyle's book of colours I once made upon myself with the hazard of my eyes . . . The manner was this. I looked a very little while upon the sun in a looking glass with my right eye & then turned my eyes into a dark corner of my chamber and then winked to observe the impression made & the circles of colours which encompassed it & how [they] decayed by degrees & at last vanished. This I repeated a second & a third time. At the third time when the phantasm of light & colours about it were almost vanished, intending my fancy upon them to see their last I found to my amazement that they began to return & by little & little to become as lively & vivid as when I had merely looked upon the sun. But when I ceased to [intend] my phansy upon them they vanished again. After this I found that as often as I went into the dark & intended my mind upon them . . . I could make the phantasm return without looking any more upon the sun. And the oftener I made it return, the more easily I could make it return again . . . And which is still stranger, though I looked upon the sun with my right eye only and not with my left eye, yet my phansy began to make the impression upon my left eye as well as upon my right . . . And now in a few hours time I had brought my eyes to such a pass that I could look upon no bright object with either eye but I saw before me, so that I dared neither write nor read but to recover the use of my eyes shut myself up in my chamber made dark for three days together & used all means to divert my imagination from the sun. For if I thought upon [Him] I presently saw this picture though I was in the dark. But . . . by keeping in the dark and employing my mind about other things I began in three or four days longer to look upon bright objects recovered them pretty well . . . But now I have been very well for many years, though I am apt to think that if I dared venture my eyes I could still make the phantasm return by the power of my phansy. This story I tell you to let you understand that if the observation related by Mr. Boyle, the man's phansy probably concurred with the impression made by the sun's light to produce that phantasm of the sun, which he constantly saw in bright objects. And so your question the cause of this phantasm involves another about the power of phansy which I must confess is . . . too hard a knot for me to untie. To place this effect in a constant motion is hard because the sun ought then [to] appear perpetually.

It seems rather . . . in a depression of the [senses] to move the imagination strongly & to be easily moved both by the imagination & by the light as often as bright objects are looked upon.

[MS, Locke c16 ff. 92–93 Letter from Molyneux (William) to the authors of the *Bibliotheque universelle,* 7 July 1688. From the Correspondence of John Locke, The John Locke Collection, Bodleian Library, Oxford University]

CHAPTER 3

MODERN STUDIES OF BLINDNESS AND TOUCH

INTRODUCTION

As we showed in the last chapter, the study of blindness and touch has its roots in the Enlightenment. In this period, it became a means of illustrating learned perceptions, a symbol of the subjectivity of memory and the experience of nature, and a mark of theological and political rebellion. In this chapter we discuss how this focus influenced more modern research into blindness, and informed the social and educational policies we discuss later in this book. These studies all use scientific definitions and classes of blindness described in the previous chapter, and tend to focus on total and congenital blindness, even though only a small minority of people registered blind fall into this category—the great majority of people registered blind in England have residual vision, and also became blind after the age of retirement (Hayhoe, 1995a).

What is most noticeable about these studies is how their attitudes to blind people evolved from those featured in the last chapter, and the development of the conclusions that they drew; from early scepticism in which morality and understanding are discussed, to the more liberal attitudes that reflected those of the twentieth century. In this chapter we also show the evolution of the more abstract, less empirical study of philosophy, and how it has often tended to continue to cling to its traditional beliefs that blind-

God, Money, and Politics, pages 37–54

ness prevents aesthetic understanding, and how it has linked perceptual and emotional concepts.

This chapter is split into three sections. The first examines how traditional research in this field evolved, and how it advanced into the psychological study of perception. The second section describes educational research and introspective studies by people with personal experience of teaching or being taught the visual arts and represents a more social psychological style of research and writing. The third discusses research into the communication of ideas, which appears to be, paradigmatically, a crossover of studies, and borrows in part from the researchers represented in the first section.

This chapter now continues with its first of these sections.

CONTEMPORARY STUDIES OF MOLYNEUX'S QUESTION TO LOCKE

The first modern psychological study modeled on Molyneux's question was conducted by Gregory and his assistant Jean Wallace in the early and mid 1960s (Gregory, 1974). This later became acknowledged as the most influential perceptual psychological study of recovery from blindness in the modern era. Gregory studied a man he named SB, who had been successfully treated for congenital cataracts at the age of fifty-two. To study this phenomenon Gregory and Wallace followed SB, observing his actions in different situations, but mainly contriving situations in which he could be tested by more structured empirical tests; these included visits to museums and city centers, in which the reactions to his first visual perceptions could be recorded. In an additional test, SB was also observed performing exercises set by Gregory. For example, it was said that, "he found that when looking down from a high window (about 30–40 feet above the ground) he thought that he could safely lower himself down by his hands" (Gregory, 1974, p. 101). In an additional exercise, when Gregory and Wallace asked SB to draw "he was most doubtful of his ability to draw, but once started, he enjoyed it, and attacked it with concentration" (Gregory, 1974, p.107).

Gregory also completed detailed notes about SB's behavior derived from his observations, with the remainder of his analysis coming from health and school records. In common with the medical definitions of blindness, he was at pains to include a treatment history, list of diagnoses and a legal definition of SB as a person who had been registered blind since birth. This was produced in much the same manner as were Berkeley's original experiments. Berkeley's original findings were confirmed by Gregory. His discussion also included a criticism of von Senden's primitive conclusions that blindness retards a sense of spatial perception. Unlike Revesz and von Senden, Gregory also suggested that totally, congenitally blind people appreciate objects aesthetically, and can discriminate discrete information, such as beauty from distasteful substances. This also made him one of the first psychologists to empirically recognize emotional judgments as pleasing and displeasing, pleasant and unpleasant. It was just that, he felt, tactile discrimination is formed from different experiences and perceptual stimuli. In his concluding remarks he wrote:

> One may even say that their attempt to see was made long before their eyes were opened to the light, and in this respect they differ not only from most other cases in the literature but also of course from infants. (Gregory, 1974, pp. 116–117)

Most revealing in Gregory's case study were his and Wallace's observations on SB's emotional state at having to change from one information culture to another, which became subsumed into his work behind his need to compare his work with Berkeley's. This need was illustrated by Gregory's emphasis on his discovery that, after a while, he appeared to transfer information from his tactile perceptions to his visual ones. He called this phenomenon Cross Modal Transfer (CMT). In his narrative, Gregory made lesser mention of the fact that SB became confused during this process, and that this affected him mentally. Although this feeling was at its strongest just after his operation, it lingered far longer than was healthy. And in the long term this clash of cultures caused by the shock of being faced with a completely new perceptual world led him to experience bouts of depression. Although SB tried to recreate a nonvisual world in private, often sitting at home and turning off all the lights, he never really became comfortable or happy in his new visual world and died shortly afterwards. Gregory found this disturbing, and finished the recording of his field notes with the following words of great regret:

> On 2 August 1960, S.B. died.
>
> His story is in some ways tragic. He suffered one of the greatest handicaps, and yet he lived with energy and enthusiasm. When his handicap was apparently swept away, as by a miracle, he lost his peace and his self respect.
>
> We may feel disappointment at a private dream come true: S.B. found disappointment with what he took to be reality. (Gregory, 1974, p. 114)

Even though institutional changes from the late 1960s until the early 1980s led to a change of philosophical direction in the treatment and attitudes toward people who were blind, scientific studies of blindness remained steadfastly linked to the methods and definitions of blindness of the Enlightenment; these studies have remained focused on a medical understanding of perception. They are now discussed below.

Studies of Blindness and Perception since Gregory

Gregory's observations were replicated in a study by Sacks. He observed a fifty-year-old man, Virgil, "who had been virtually blind since early childhood. He had thick cataracts, and was said also to have retinitis pigmentosa, a hereditary condition that slowly but implacably eats away at the retinas . . . " (Sacks, 1993, p. 63) but recovered his sight after an operation. In his study, Sacks took an approach as empirical and detailed as Gregory's when describing the legal and medical nature of Virgil's blindness, and consulted his predecessor on the nature of his experiments. In addition, he was also given permission to refer to the diaries of Virgil's sister, collected from childhood to the time of the study.

In terms of testing Virgil, Sacks discovered that his examinations of medical records provided little usable or objective results, and thus he also questioned the validity of previous medically based case studies analyzed by von Senden. Perhaps unsurprisingly, in his narrative he felt that Gregory's methodological tools and models of analysis were more useful, and similarly copied the structure of his analysis, ultimately writing a thorough, narrative case study. For example, after first meeting the subject at the airport, Sacks described "[Virgil pointing] to all of the cars we passed . . . 'Look at that one' he exclaimed once. 'I have to look down!' and bending felt it" (Sacks, 1993, p. 66).

In the narrative, Sacks's findings largely confirmed Gregory's earlier results, and cast further doubt on von Senden and Revesz's more primitive belief that people who had never seen could never appreciate beauty. Interestingly, he also picked up on Gregory's earlier observation on the social and emotional development of blindness. He noted that Virgil had similar difficulties adjusting to a different perceptual culture, causing him to become depressed and recreate his earlier blindness. This suggested to Sacks that Virgil had indeed had a different set of expressive references when perceiving his surrounding environment that affected him emotionally. However, in a strange twist of fortune, a further event made this observation of little consequence in the long term. Following a further illness Virgil re-lost sight shortly afterward. This led Sacks to write:

> [Now] a final blindness—a blindness he received as a gift. Now, at last, Virgil is allowed to see, allowed to escape from the glaring and confusing world of sight and space, and to return to his own being, the touch world that has been his home for almost fifty years. (Sacks, 1993, p. 73)

Since Gregory and Wallace's study, more laboratory-based psychologists have also used a range of scientific and quasi-medical methods. These have broadly supported their argument that powerful and meaningful information, in many cases comparable to that of sight, and an understanding of *visual* aesthetics, are accessible to people who are blind. In a study of spatial awareness through active touch, for example, the British educationalist W. D. Jones (1970) compared the tactile perception of several sighted, early and congenitally blind school children, their ages ranging from five to twelve. Although he supported a relatively loose definition of blindness, and his experimental methods did seem, on reflection, to be far less rigid, structured or ethical than Gregory's study—Jones defined blindness as "at most perception [of] light" (Jones, 1970, p. 37)—his results showed a more sophisticated understanding of the differences in aesthetic spatial perception when derived purely from nonvisual means.

During his study, Jones blindfolded sighted children, and compared them by age to children who were blind, which he felt would create a fairer developmental comparison. Jones then pressed the tip of a pencil onto the arms of the children, and they attempted to mark the spot where they had just been touched with their own pencils. The difference was measured and his results were presented numerically. In his analysis of the data, Jones argued that "although the blind children are inferior to sighted children in complex shape discriminations, they are more accurate than the sighted in terms of [the] basic tactile . . . sensations involved" (Jones, 1970). This suggested that the children who were early and congenitally blind derived more accurate spatial information through their skin, although the shape of the objects was better understood when viewed *as a whole* through sight.

Comparable conclusions were made by the British psychologists Jane Dobson (1982) and Alan Dodds (1980), who conducted structured experiments to study the spatial awareness of students who were congenitally or early blind and those who were sighted under laboratory conditions through touch. In their report of the study, they argued that awareness existed in both groups of people that cannot be accounted for by using the traditional frameworks and analysis of understanding the dichotomy between sight and touch usually associated with traditional studies of perception, such as that of von Senden. Dodds wrote in his conclusion:

> [von] Senden's view that the congenitally blind cannot represent space must now be laid to rest. [The] findings that at least one congenitally blind subject could achieve such a feat [understand spatial perception to suggest they understood perspective] leads to a fairly firm rejection of such a notion. (Dodds, 1980, p. 33)

These more advanced conclusions were claimed in an era that was beginning to understand the ability of students who are blind in government, and the development of a disability movement that was gathering credence in many Western societies and entering mainstream politics. However, even in this atmosphere of social progress, scientific definitions and their underlying philosophies remained rooted in the notion of blindness as a reductionist, abstract, cultureless phenomenon, in much the same way as it was in the Enlightenment. Moreover, blindness remained a phenomenon that would define a class of people and behaviors, rather than as a trait that was one among many others, and that defined a whole individual.

More important, this method of investigation still influenced the majority of modern writers on blindness and the interpretation of tactile perception after the legal integration of blind students into English schools. In addition, these writers continued to focus on people who were blind as a group of disabled figures on the fringes of human understanding, and still needing therapy; even if this therapy was now psychological rather than medicinal. This view affected the attitudes of those in positions of power over institutions (Hayhoe, 1995a, 2000).

This chapter now continues by discussing modern philosophical studies of blindness and nonvisual perception, as a way of illustrating a more contemplative analysis of blindness, touch and the classification of people who are blind.

Contemporary Philosophical Studies on the Nature of Perception

There are sweeping differences between contemporary philosophers in their thinking on the ability of people who are blind. These follow the pattern of the analytical methods used and the subjects of analysis, suggesting two distinctive schools of thought. Two contemporary essays by the sighted British and American philosophers Bryan Magee (Magee & Milligan, 1998) and Thomas Nagel (1991) relied on the methodology of introspection. In the first, Magee discussed the blindness of his fellow British philosopher, Martin Milligan, in open letters between the two, much in the style of John Locke and his contemporaries almost 300 years before. Nagel, on the other hand, conducted a more existential study, asking whether it was actually possible to imagine the lived reality of another creature, his main point of reference being a bat, which had a completely different frame of reference and information culture in its consciousness.

In their analyses, both authors doubted whether people who have never seen can ever have a true sense of aesthetic appreciation of objects, or understand corporeal qualities that can only be wholly understood by vision. Their opinions on this concept lend a great deal to a pre-Lockeian notion that objects have intrinsic characteristics that make themselves understood by people with access to such a faculty. In this way, to understand, or to have an emotional or aesthetic connection or knowledge of such an object, you have to be able to sense it in the way it was designed to be perceived. Nagel especially felt that it was not possible to describe a color by using sound analogies, as he had previously experienced.

Similarly the British philosopher Robert Hopkins (2003) reviewed psychological and philosophical literature on the perception of perspective and aesthetic beauty, a theme of understanding in the philosophy of aesthetics that also owes a great deal to Locke's notion of blindness as an object of understanding. More contentiously he also used introspective methods of understanding blindness in this context. Broadly agreeing with Nagel and Magee, Hopkins also considered aesthetic understanding to be based on a comprehension of the media being observed, rather than a collation of information that could be passed between the senses. In the fine arts, he argued that only sculpture can be appreciated for its beauty without vision, whereas two-dimensional art works appear inaccessible to people who are blind.

More recent writers who have benefitted from a broader literature have disagreed quite strongly with these conclusions. The Irish philosopher David Feeney (2004), who analyzed literature on the theatrical arts among others, and the British art theorist Bruno del Tufo (2005), who reviewed and observed the work of artists and students who were blind first hand, found that the vocabulary and understanding of people who are blind and have observed all forms of artefacts were as rich and mature as those of people who were sighted. This, they concluded, implied a developed understanding of their whole environment, one that transcended a single perception, and more one that included many perceptions that could combine information to form a holistic picture.

This body of knowledge is not the only one to focus on the nonvisual appreciation and understanding of objects based on earlier philosophical debates. This theme was also investigated by educational psychologists who have tried to relate this knowledge to different branches of education and the role of the senses in pedagogy. These studies fall into two categories: studies of three-dimensional aesthetics/spatial understanding and blindness, and studies of *two-dimensional* awareness and blindness.

A discussion of these studies is presented in the following two segments of this chapter.

STUDIES OF THREE DIMENSIONAL AESTHETICS/SPATIAL AWARENESS AND BLINDNESS

Previous surveys of this field (Hayhoe 2003) indicate that arts theorists and educationalists have considered three dimensional modeling to be the main, and in some instances the only form of art education accessible to people who are blind. More to the point, only a few theorists in Europe and the United States accept that some two-dimensional representation is wholly accessible to students who are blind (De Coster & Loots, 2004; Harris, 1979; Mülheis, 2004). As a result previous educational models favored three-dimensional art education models, based on haptic—or active—or proprioceptive—often referred to as passive or intramuscular—touch. These models have also been informed by the work of the earlier era of psychological study—including revisiting Revesz, von Senden and Gregory—but have more recently led to a separate theory on the nature of learning activity.

Although research on blindness and learning has appeared to determine a post-Reveszian era after Gregory's study in the 1960s, writers have continued to use his model to study the three-dimensional representation of models by children who were blind under laboratory conditions (Kinsbourne & Lempert, 1980; Witkin et al., 1968). These are important to our development of an understanding of the perception of three dimensions. However, studies of the understanding and performance of students who are blind in classroom settings, and those focused on the use of touch and modeling as a form of emotional and intellectual development, have provided the most valuable insights. In particular they have not only provided an understanding of the nature of perceptual awareness, they have also examined the social and historical settings of this work.

This group of studies are now discussed below.

Studies on Emotional Development and Three Dimensional Modeling

The original focus of study in this field was three dimensional modeling as a form of language, which became similar to a version of tactile language. This was a very different focus from the parallel studies that existed in the same era as von Senden, which focused on perception as a channel of aesthetic appreciation, and received much less attention by the wider research community. The first educator to study this field in the twentieth century was the Austrian educator Viktor Lowenfeld who, after training as a sculptor, worked in a Viennese school for the blind in the early 1930s. During this post he wrote his first case study about a student there, a congenitally blind and deaf girl called Camilla, whom he taught how to sculpt in between her other classes. This was thought almost impossible by the other teachers in the school at the time (Micheal, 1981).

In his description of their first exercise, Lowenfeld writes about how he formed a model of a human body from clay coils. After making this model, he attempted to communicate her body parts through first touching the model with her, and then touching her body part to the part of the model it was associated with. At first this method did not work, and led to a great deal of frustration on both of their parts. Nevertheless, after a while she began to show definite signs of understanding, which Lowenfeld found later developed into a sense of aesthetic understanding and tactile communication. In his narrative of the study, Lowenfeld described how through this steadily evolving system of lessons he could also build on Camilla's confidence and give her an expressive outlet not present before. This was a revelation, as before he began this educational experiment, he found that Camilla rarely communicated with the other staff or students, and was often unruly. The beginning of this process is particularly well described in an early passage:

> Then one day I had her sitting close to me. I started with the thick coil again. And this time I went over her body again and then took her hand and let her touch the thick coil. . . . All of a sudden she shouted "Eeeeee!" and she grabbed the clay. For the very first time I had the feeling that we had both communicated. (Micheal, 1981, pp. 12–13)

After moving to the United States as World War II approached, Lowenfeld continued his educational experimentation and research with children, both sighted and blind. During this time his methods became based more upon artistic objects rather than their creators—this has been referred to as object-based research—and he presented fewer case studies of individuals. Instead, in this period of work he created many more structured scientific observations that were constructed around large studies of many students (Lowenfeld, 1959; Lowenfeld & Brittain, 1987). In this period he also became interested not only with children who were blind, but increasingly those with partial sight as well. From these later studies his writing suggests that he became more aware of studies by those such as Revesz, as he used more empirically based models of haptic[1] and visual types to describe two classifications of children and their making habits.

Another aspect of Lowenfeld's research is little considered, but is arguably just as important. As he began this more pioneering form of narrative on the experience of blindness in schools, he concentrated on developmental definitions of blindness, rather than visual acuity alone. These were based on individual aesthetic appreciation rather than levels of visual perception. This was a theme that was also to be addressed in more detail by his brother, Berthold Lowenfeld (1981). Viktor Lowenfeld's belief in the importance of understanding of emotional as well as physical and cognitive development was later described in my ethnographic studies of art classes during the 1990s and around the turn of the millennium (Hayhoe 1995a,b, 2000, 2003, 2005a). In these narratives, I found that aesthetic appreciation was developed by individual personality types, based on factors such as those described by Warren (1994), rather than levels of perception.

Our observations have also recently been substantiated in a review of case studies by Sacks (2003), entitled *The Mind's Eye.* In this work, he describes how an appreciation of the *need* to touch by authors who are blind, like the need to see, has led to a finer, more introspective comprehension of their world. In his review Sacks found that a sense of aesthetics was more culturally important to the totally blind authors such as Hull (2003) than it was to the sighted, as it allowed them to forge a sense of cultural identity as they come to terms with their blindness. In this way, people who are blind feel that the communication of cultural symbols is of the utmost importance to their sense of social identity, which is often stolen by other means. This concept is illustrated through an excerpt from the work of Benson and McNeice:

> We have learned much through our hands, through our tactile comprehensions, and proprioceptive deep muscle sensitivities. We truly comprehend the meaning of hundreds upon hundreds of word concepts only because we have literally felt the hot stove, the cold ice, the sharp edge of a knife. (Benson & McNeice, 1964, p. 46)

Sacks also found that the minds of people who are blind chose how to distinguish beauty and sensations that it felt pleasant, as well as choosing to preserve visual memories or to reject them in favor of a new, sightless, emotional world. In his text he compared the fundamentally different imaginative experiences of Hull to those of the Turkish author Torey. Torey felt that his visual memories were enhanced because he had become blind. Hull on the other hand found that he lost all sense of what it is to see, and only committed tactile thoughts with his previous memories, even though he had become blind as an adult. Both men's minds, Sacks continued, had therefore chosen a course for their imaginations based on their individual experiences before and since becoming blind. This brought into question the scope of other sensual perceptions in creating emotional images of artefacts or environments. As he wrote in his conclusion:

> [One] can no longer say of one's mental landscape what is visual, what is auditory, what is image, what is language, what is intellectual, what is emotional—they are all fused together and imbued with our own individual perspectives and values. (Sacks, 2003, p. 59)

Surprisingly Sack's observations have not been discussed widely by other authors in this field, and his findings are regarded as esoteric by some psychologists, educationalists, art theorists and philosophers—perhaps this is because his finding were difficult to replicate using experimental methods, which allows an easier scientific interpretation of blindness. Thus, there is a limited amount of empirical research in the field of commercial psychology and education.

This section continues by discussing studies of the nature of touch and creativity.

The Role of Touch in Creativity

Regarding the role of touch in the creation of a broader image of sensations, an area not widely studied, evidence shows some interplay between our perceptual understanding of the world around us, and a certain element of creativity in the way we perceive our surroundings. Experimental studies, however, are relatively new to this field and so have not provided a broad field of knowledge. Perhaps the best known of these modern studies is the Secrets of the Senses project at Oxford University (Spence & ICI, 2003), directed by the British psychologist Charles Spence. His team have conducted laboratory experiments on individuals stimulated by several differing perceptual stimuli simultaneously. Spence's initial findings suggested that the perception of one sense, such as vision or hearing, can be changed by changing other perceptual stimuli, such as smell or taste.

Although employing less rigorous methods, definitions and executions, an earlier study by the Senior Master of the Royal Liverpool School for the Blind (RLSB, 1970) in England also hinted that the verbal representation of nonvisual images also showed a complex picture of emotional and cognitive metaphors in his students which related one sense to another. In his experiment the Senior Master compiled a list of words he considered unrepresentative of visual concepts. These included words such as "buzz," "drum" and "deafening." A group of blind children of different ages and IQs were then asked if they could comprehend these words, and a list of the results were then compiled as rhetorical comments, from the most comprehensible at the top to the least at the bottom. After conducting these tests the Senior Master concluded that experience had allowed students to cross-reference words they had experienced with words that were often associated with more visual concepts, and thus had developed a complex map of mental associations and links.

> [The] liveliness of the children's response convinced me that the vocabularies of sound, touch, smell, and taste, even though together only equal to the vocabulary of sight, are formed by a common and rich interpretation of experience. (RLSB, 1970, p. 58)

These studies seem to agree with Gregory's earlier observation that cross modal transfer between touch and vision allowed our brains to pass some form of perceptual information to the interpretation usually employed by another's. More important, these findings noticeably contradicted the philosophical studies of Nagel and Magee, which doubted that blind people have the ability to understand what are considered to be nonvisual concepts. Although experience of these *perceptually holistic* concepts have reason to suggest that people who are blind have developed a creative sense of their environment through their sense of touch, studies which examine the understanding of two-dimensional concepts appear to suggest a more complex, controversial debate. However, this model of academic development still appears to be based on traditional cognitive reasoning.

It is to a review of these studies that this chapter now turns.

STUDIES OF TWO-DIMENSIONAL AWARENESS AND BLINDNESS

The Swedish historian of tactile images Yvonne Eriksson (2003) notes that a large number of studies examining the perception of pictures through touch exist. Within this corpus of research the most prevalent factor is the cognitive process of converting three-dimensional artifacts and environments into two-dimensional representations. The vast majority of these studies are again undertaken in highly controlled laboratory conditions and use medical and legal definitions of blindness. They are also mainly still interested in perceptual comparisons alone. For instance, many researchers compare the work of children with sight and that of those who are blind.[2] The most common current debate, however, is that on the inherent ability to draw and paint.

The Anglo Canadian psychologist John Kennedy's (1997) review of his first era of research in his book Drawing and the Blind examined the tactile capability of people who were blind to comprehend two-dimensional pictures. His early experiments featured haptic drawings that those involved in his observations touched or drew. These showed that his research subjects could perceive elements of two-dimensional drawings which were previously only thought to be visual through touch alone. Kennedy (1983; Kennedy & Merkas, 2000) has also found that by comparing ancient drawings, such as cave paintings, to those of people who are blind he observed similar elements, suggesting that there are parallels in their untutored state that were unaffected by an all imposing visual culture. In his later conclusions, Kennedy (1997) proposed that a two-dimensional language can be established that is understandable to all people through tactile graphics.

Findings similar to Kennedy's have been established by the British educational psychologist Ron Hinton (1991), in studies under laboratory conditions contrasting work perceived by sighted and blind students. In the excerpt below, Hinton describes how he found that comparing raised line drawings by people with various levels of vision appeared to show a similar understanding of certain tactile cues. These suggested a commonality of perceptual experience:

> In an outline drawing, key facts about the sizes, shapes and locations of objects can be shown in a form of representation that requires less training than any code or language, less time than listening to a description, and little if any translation of its message to be universally understood. (Hinton, 1991, p. 82)

More controversially among the traditional research community, Kennedy has also proposed that visual concepts, such as perspective and visual metaphors, can be understood through tactile perception alone. By comparing the comprehension of drawing by early and congenitally blind, and sighted children, he discovered that children who were blind not only had a basic sense of foreshortening—the ability to understand that objects become smaller as they fall into the distance—but also recognized raised symbols meant to suggest distant movements, such as speed lines in a cartoon (Kennedy, 1993, 1997; Kennedy & Merkas, 2000). From this he concluded that there was also a universal understanding of even extreme visual representations of distance and symbolism. It is also Kennedy's view that this new corpus is based on a concept in which its participants are motivated to *work toward* the goal of understanding the likeness of perceptual ability—a scientific, experimentally testable stance—and, although in contrast to his own theories, allows for the historical and social context to be considered.

Kennedy's findings are strongly disputed by authors such as Hopkins (2000), who contends that people who are blind have the ability to *learn* that distance leads to a narrowing of objects, and have a simple understanding of immediate distance. This occurs when reaching forward into the distance. According to Hopkins this ability is learned through more simple mechanisms than those required in visual foreshortening, and therefore the cognitive functions that produce this level of understanding differ markedly among people who have always had sight and those who have never seen.

Despite disagreements about the holistic nature of perception and cognition, Kennedy finds that the body of knowledge that is finally developing in this field is taking the form of an identifiable *bio-social* genre of research on blindness, with its own methods and narrative. This can draw from the varying debates on the differences of reproduction of sighted and blind experiences. This he discussed in an email to Hopkins, with whom he continued a personal correspondence. In his email Kennedy felt there can be agreement that there are elements of perception all people, no matter what their disability, can understand. Furthermore, he contended that academic knowledge about pictures can be seen as more objective than the attitudes shown by institutions in this field to people who are blind, as they rely less on detrimental personal interest and the *psychology of images* is in itself less political than blindness. In this email he emphasized:

> The basis of this work is a theory of major universals in vision and touch, such as line depicting edges of surfaces, plus a child-developmental sequence carrying both the blind and the sighted from scribbles to some perspective devices, plus a cultural learned overlay. So limits on the blind and theory of perception that were set fast for centuries have been moved far a-field. New limits are hypothesized by Kennedy (line does not show shadow borders for reasons explored by Cavanagh) and Hopkins (though surely perspective can give direction impressions, he argues perspective is unlikely to give depth impressions in touch). Relevant pathways in visual brain are found by Sathian and Pascual Leone.
>
> The referents and representations are multimodal, in ways explored by Spence. Skilled knowledge in one modality (such as athletic prowess) is trapped, and we need to aim to free it to be paraphrased in other modalities, and deepen it. You may throw a ball well, but have no ability to describe momentum. One step beyond throwing is knowledge enough to coach, but a big step will allow you to send a rocket to the moon. You may jump just fine, but not be able to describe gravity at all.
>
> Previous conceptions of the blind were driven by social factors, Hayhoe tells us. A theory of pictures, representation and knowledge is more general and may be immune to winds of policy. [Personal correspondence from John Kennedy to Robert Hopkins, in summary of a teleconference on Thursday 16th October 2003, and sent on Thursday 16th October 2003. Reproduced with kind permission from Professor John Kennedy]

We now conclude this chapter.

CONCLUSION

Although not apparently influenced directly by religion and political thought as it was in the past, the modern study of blindness and the senses has a complex history. What is also clear is that the definitions and strands of understanding that were present in the past, are still with us today in the extreme understanding of blindness as the total loss of sight, and of blind people as having no conception of what it is like to see. In the twentieth century this led to many different corpora of theory on development growing, from an abstract understanding of perception through experimental environments in which real situations were vaguely simulated. This understanding led to the belief that the perception that blind people relied upon most was touch. Like their earlier counterparts, modern academics who have studied blindness have also often regarded it as almost allegorical of defective human thought that can expose greater truths about the whole of human perception, rather than the education or greater understanding of the blind as a cultural, social or institutionalized community.

These issues are now discussed under separate subheadings.

Reduction, Abstraction and Methods in the Study of Blindness and Touch

The study of blindness and touch in the modern era often remains the preserve of abstract philosophers, a small number of educationalists and, increasingly, experimental psychologists. These studies have inherited an abstract approach to the definition of blindness, an obsession with medical and legal definitions, and an understanding of thought that almost exclusively regards touch as the only nonvisual method of gaining education, understanding or conducting an emotional relationship with their environment. This is very different to the lived reality of people who are blind, as it takes no account of the other remaining senses in ordinary life, let alone, as Spence explains, the interplay of the senses in building a more complete mental image.

Therefore it is often the case that research in this field is not rationalized from the point of view of working toward an understanding of blind people, or even the lived reality of blindness. It studies blindness from a wholly abstract notion of what it is as a laboratory condition, described in order to pursue an understanding that is palatable and comprehensible to the academic community. It does not account for other human factors on understanding, or see the thought processes of people who are blind in the same social realms as those who are sighted. In this way, the concept of the *blind person* becomes that of *the other*, an abstract species that does not exist in our world, but some form of parallel world.

Another problem with modern research into blindness is the method which it employs for two overarching reasons. First, the nature of science itself is to reduce concepts that, despite their diversity, have a thin connection, and corral them into a single category so that they can be classified more easily. Second, scientific exploration needs some form of historical framework from which to work. Studies are too infrequently generated from the needs of a human community, rather than the needs of a more abstract *human understanding*. The speed of evolution in scientific thought, as Kuhn (1996) suggested, is therefore restricted by the paradigms that bound its operation and freedom to experiment with new ideas. Our review has shown that the study of blindness follows this fashion.

A consequence of empirical methodology is that analyses are held within straight jackets. Academic departments which study blindness and their systems do not generally cross-fertilize, and a need to adhere to the letter of the empirical method suffocates new forms of intellectual creativity. As Lowenfeld found, people with whom he taught in Austria and who were raised on science, were more likely to follow a generally held belief that the students in his school were intellectually incapable. As Kennedy found, a broader depiction of tactile picture perception, taking into account medical, philosophical, psychological and social factors, is only now starting to be understood after centuries of research, and only after its break with a conception of the study of *only* blindness and touch. It is thus that science too often reduces naturally diverse phenomena, such as a lack of sight, into an instantly compact concept, understood, diagnosed and studied through a minimal and oversimplified concept such as blindness.

The Study of Total Blindness

Another legacy of Locke and his colleagues is the almost total concentration on total blindness in the modern era. This historical legacy, however, cannot only be seen as a need to reduce definitions. If this were the case, it would be logical to choose the most economically debilitating and common form of visual impairment as indicative of blindness, and build a definition based on it. This would suit the reductionist tendencies of academia and governments simultaneously. It would also satisfy the financial needs of charity.

To chose the focus of total blindness and touch that is studied alone in the great majority of research, indicates the continuation of Locke's academic need to use blindness as an anomaly that can help us understand the broader human condition, not just that of people who are blind. This would also go some way to explaining why research often *compares* people who are wholly early blind to those with sight, as these represent the most extreme examples of the human categorization of perception; these people have never seen, and thus visual conceptions are a closed world they are controlled by but cannot grasp.

Perhaps this then is the underlying incentive for the greater academic hunger to target blindness above all other disabilities. This focus of research reveals a deeper element in the psyche of contemporary researchers into the use of the senses. Blindness is above all an information disability. Smell and taste cannot convey higher meaning. These senses are primitive and possess no logic. Moreover, even though it is infinitely complex and sophisticated, sound is not a reliable information source. It is temporary, transient, and open to distortion within time and over the passage of time.

Thus it is not trusted as a reliable source of storing wisdom, knowledge, aesthetic insight, or metaphysical scriptures that are felt to contain forms of spirituality within their very symbols; a sound can never be an icon.

For this reason, deafness or the inability to taste or smell cannot be seen as an allegory of the incomprehension of higher truths. Deaf people can read given the opportunity. They can see paintings. They can interpret graphs and diagrams. They can kiss icons, and understand their form and meaning. It is often thought that they may not find it easy to tell people what they are feeling if they are born deaf —thus it was felt that those who could not communicate their thinking verbally were said to be dumb, a term also used for a lack of intelligence. But they can be touched by the importance of this visual information.

Above all disabilities, therefore, early total blindness lies alone in the realm of human fascination as a disability that not only impinges on perception and a reading of intelligence, but one that cuts to the heart of the *human soul*. For many researchers, to understand what blindness lacks is to comprehend what vision provides: beauty, knowledge, understanding, compassion, civilization and even God. For them, to provide this to a person who cannot understand these things is the noblest task of all.

Although this book has focused critically on what it considers to be many of the negative aspects of the study of blindness and touch, there are many elements of this type of research that are often helpful in the education of students. They represent a more contemporary model of research and writing in the field. Although these researchers were and are fascinated by the structure of thought and how their general understanding can add to knowledge, their explorations have political as well as practical and humanitarian motives; these have been interpreted through this lens and shaped our attitudes toward people who are blind or visually impaired.

It is to a search for new models in England and their effects that this book now turns. In doing this we discuss the contexts provided by the academic research in this field outside of the laboratory and the philosopher's study, and investigate why, if as Kennedy, Sacks, Jones and Spence are to be believed, touch is not the only form of examination we should be investigating, even though centuries of research and debate have focused almost solely on it.

In the following part of this book, we investigate whether these definitions and the structure of our understanding of blindness and touch, its classifications and definitions, have helped or hindered the lived reality of classes for students who are blind. In particular, this book focuses on the cultural influences of a *political* manipulation of blindness, and how this has affected an understanding of disability in England over the past two centuries.

We begin this process in the following chapter by describing the foundations of the first asylums and schools for the blind in England.

NOTES

1. Haptic children are those that refer to nonvisual cues when making art works.
2. See Gabias (2003), Jansson and Holmes (2003), Heller (2003), and Kennedy (1974, 1993, 1997), Millar (1975).

PART II

CHAPTER 4

VOCATIONAL EDUCATION IN ENGLISH SCHOOLS FOR THE BLIND

In Part Two of this book, we discuss the development of the vocational curriculum in English schools for the blind, and the changes in attitude, culture and politics toward blindness and touch. Rather than isolating this discussion merely to English politicians, educators, administrators and theorists, however, Part Two also discusses the influences of other countries in the process of this cultural evolution. This chapter begins this discussion by examining the cultural attitudes that influenced the education of the blind in England, and the social factors that affected them. This analysis is split into the following three sections: the introduction of education of the blind into England, the role of handcraft in the education of the blind and a summary of this chapter.

THE INTRODUCTION OF EDUCATION OF THE BLIND IN ENGLAND

The first two English asylums represent the second (Liverpool, 1791) and fourth (Bristol, 1793) institutions in the world for the education of the blind, and their vocational curriculum has been copied by asylums, which were introduced in England after this period until the second half of the nineteenth century. They were predated by the Paris Institute, conceived in

God, Money, and Politics, pages 57–75

its earliest form in 1784 (Illingworth, 1910), and Bristol was predated by the Edinburgh Asylum in early 1793.

The first formal proposal for segregated education of the blind was published in an open letter to the *Edinburgh Magazine and Review* in 1774, by an anonymous author signing himself only Demodocus (Demodocus, 1774). In this letter the author drew from the approach of the French philosopher Denis Diderot,[1] the English Mathematician Nicholas Sanderson (who was himself blind and sponsored at Cambridge University by Isaac Newton (Del Tufo, 2005), Berkeley, and Locke's earlier debate with Molyneux.

Drawing on examples of perceptual studies presented in Diderot's (2001) *Letter on the Blind*, Demodocus, who had lost his sight, observed the increasing amount of beggars who were blind and unable to find a place in society. In the process of this argument, and unlike the studies at the beginning of the Enlightenment that had treated blindness as a more hypothetical phenomenon, he attempted to elevate the cause of the blind population to a human cause. Demodocus justified his case by presenting examples of people who were blind and successful in the academic and social fields.

In his essay, Demodocus reflected Diderot's earlier contentions that blind people could often be regarded as victims, whose ability was misunderstood and whose circumstances could only be rectified through a balanced system of education. This, he felt, would allow the students to be both enlightened intellectually and provided a trade at their leaving age. This would lessen their chances of poverty later in life. The excerpt from the letter below illustrates this element of his conclusion.

> [In the past the] blind were objects of compassion, because their spheres of action and observation were limited. . . . The most important view, therefore, which we can entertain in the education of the person deprived of sight, is to redress, as effectually as possible, the natural disadvantages with which he (sic.) is encumbered; or, in other words, to enlarge as far as possible his sphere of knowledge and activity. This can only be done by the improvement of his intellectual imagination and mechanical powers, and which of these ought to be most assiduously cultivated, the genius of every individual alone can determine. . . . That if one sense should be suppressed, it but retires into the rest. (Demodocus, 1774, p. 676)

Demodocus' letter was followed twelve years later with the publication of an essay by Valentin Hauy (1889), the French translator, educationalist and social reformer, who two years earlier founded the original Paris Institute—although at this point it was just a small school in his house. Using similar sources of intellectual inspiration, Hauy also drew upon the earlier works of John Locke, George Berkeley and Denis Diderot in his essay.

Hauy's essay, however, differed from Demodocus' in terms of its emphasis. His essay rhetorically draws more from Diderot's ideas of the moral purity of people who are blind, rather than the scientific nature of perception. According to Paulson (1987) this emphasis led to the pursuance of education as a political moral exercise as well as an intellectual one. At the same time, it was also an important factor in the history of the education of the blind that Hauy dedicated his essay to King Louis XVI, then the king of France. According to Jay (1994) and Paulson, this pursuance of the King's approval was Hauy's attempt to secure royal patronage for his institute and therefore ultimately led to the linking of the first institutes for the blind in continental Europe with other popular royal causes and charitable works. In Britain, however, this notion of charitable foundation appeared to differ. It is to this notion that this chapter now turns.

The Influences on the First English Asylums

Although drawing from the intellectual experiences of Hauy's institute in their early literature (see Hayhoe, 1995a,b, 2000; Illingworth, 1910), English institutions were created as part of a greater asylum movement, aimed at reducing a plethora of social problems, from madness to poverty. These were founded by Protestant foundations without government influence or funding. English and also Scottish port cities such as Liverpool, Bristol and Edinburgh were to be key locations for developing the first asylums for the blind, and these initial settings had an influence on the attitudes held toward people who were blind and their curricula, primarily for two reasons.

The first reason was financial. These ports were in wealthy cities. They traded commodities such as slaves, sugar, chocolate and cotton, and exported home-grown commodities such as coal, iron and wool, providing substantial profit and employment. This would have allowed the cities' merchants surplus funds with which to finance projects such as asylums. In turn, the wealth of these port cities attracted blind beggars who knew they were more likely to earn a more lucrative living from those with extra funds to spare (Porter, 1997). This was particularly noticeable in the original student records of the asylums. For example, the register from the Bristol Asylum, beginning in 1793, describes students from other areas of the West Country, Britain and Ireland lodged at the asylum, after being taken in from the streets.

The second reason is related to the causes of blindness in the eighteenth century. As they are today, port cities would have been more likely to have carried the diseases which caused blindness, such as syphilis—it is thought this could have led to a sense that blind people were immoral and sexually deviant (Del Tufo, 2005). According to the British historian Roy Porter (1997), the first European syphilis epidemic began during the conflict between French and Spanish armies in the region that would become Italy during the fifteenth century. In his book *The Greatest Benefit to Mankind*, he described how prostitutes followed these two armies' camps[2] and then moved on to have sex with soldiers in further areas of conflict and encampments throughout Europe. This epidemic spread across borders and through ports over the course of the following centuries. By the end of the eighteenth century, as Porter suggested, syphilis had become rife. It would therefore have been a major cause of blindness and mental illness in Britain and other European countries. This changed little until the widespread use of antibiotics in the twentieth century (Stanier et al., 1987; Scholl in Del Tufo, 2005).[3]

The large port cities and industrial areas that hosted asylums would therefore have sizeable populations of sexually active single men in the military and merchant navies, which were far away from their wives and mistresses, and sizeable populations of homosexuals.[4] Thus, the likelihood of contracting diseases from a range of other countries in these cities was increased. This chapter now continues by examining evidence about the influence of these factors on the curricula of the asylums.

THE DEVELOPMENT OF CURRICULA IN THE ASYLUMS

The earliest asylums in England felt that their students should only learn handcrafts, or what they sometimes referred to as the industrial arts, all day long (Bristol Asylum, 1800; Hayhoe, 1995a,b, 2000; Howe, 1833a; le Cue, 1992).These handcrafts were based on simple, repetitive tasks, which did not demand intellectual skills but a significant amount of close, intricate skills, such as cane weaving baskets and chairs by men, and simple cross stitch by women. This form of handcraft in particular was derived from three principles (Hayhoe, 1995a, 2005a): morality in behavior, the abandonment of indolence and a need to generate funds. Culturally, English asylums were distinctive from their counterparts on the Continent.

Table 4.1 illustrates the main differences between English asylums and what evidence indicates are the two other influential countries in this field. It shows the main functions of these institutions through their names, their educational aims, the methods they used to achieve these aims, the motivations that impelled the creation of these institutions, the cultural inspiration that induced the creation of the institution, what happened after these schools were created, and finally the sources of funding that kept the institutions going.

The English Protestant founders created asylums according to two fundamental tenants of their religious beliefs. First, they were formed on the idea of charitable donation, and so importantly without government interference. The second and most important tenant is the concentration

TABLE 4.1 Classification of Attitudes to Education of the Blind[a]

	French	British	Austrian
Name	Institute	Asylum	Institute
Product	Knowledge	Capital/Hand crafts	Emotional Development
Process	Route Literature	Mechanical Production	Aesthetic Hand Usage
Motivation	Morality	Morality	Morality
Inspiration	Catholic Enlightenment	Protestantism	Catholicism
Outcome	Financial Weakness	Financial Survival	Financial expedience
Funding	State, Royal Patronage, Charity	Charity	State, Royal Patronage

[a] Adapted from Hayhoe (2006).

on handcrafts and work as a form of savior. As the German social theorist Max Weber (2001) finds in his book The Protestant Ethic and the Spirit of Capitalism, populations of Protestants in many Western countries have long believed that hard work and the pursuit of an *ethical* capital will help an individual to attain an understanding of and acceptance of God.

Hence Weber noted that many nineteenth-century cultures where there were a large number of Protestants often showed a distinct gap in the wealth and professional success between Protestant and Catholic workers. Thus in asylums controlled by English Protestant foundations, as a mirror of the industrial revolution in the surrounding society, the principal method of salvation for *indolent* beggars was hard, capital producing handcrafts. The details of these principles are now investigated below.

The Role of Morality in Behavioral Control, the Division of the Curriculum and Working Hours in England

The Bristol asylum was founded by a Quaker influenced foundation led by two local wealthy and important philanthropists, Bath and Fox, and the Liverpudlian asylum was founded with the help of wealthy Unitarian and Anglican socialites and church ministers. In Scotland a similar pattern existed in Edinburgh, whose institution was founded by a Presbyterian foundation, some of whom were blind, and included a Presbyterian cleric (RBAS, 1993). In England, in particular, this uniquely influenced each asylum's moral outlook.

As an illustration of this, a poem from the report of the Bristol Asylum at the end of the eighteenth century and an excerpt from Bonners' Journal, a Bristolian periodical from 1792 describing the foundation of the asylum, explained that the founders considered it deplorable that people who were blind and *stricken* by poverty were to be found in a Christian Bristol community. They should, it was said, be pitied and helped from their mentally and socially depraved condition. The poem in particular suggested that these people, when observed in this position, were *useless.* The poem continued that as a consequence of their lack of sight, blind people already in the city could not examine significant *works of nature or art,* which many Protestants believed would enrich their lives and allow them to see the wonders of God. As a result, these founders decided that these unfortunate beggars had to work hard in repetitive labor in order to qualify for the morality that was otherwise denied to them. Below is an excerpt from this poem:

Then view you pensive, interesting group,
Hard is their lot,—with poverty they stoop,
The shades of darkness on their eyelids dwell,
They know not how to chase the mystic spell.

View nature's soul! doth not that god of day!
Pour in delight upon the visual ray?-
View Flora's beauties in their gay attire,
Say, do not these a secret joy inspire?

The charms of nature, and the works of art,
To different minds their various joys impart;
Save where the darken'd optics ne'er could learn,
Objects of arts, or nature to discern.

This night of nature striving to illume
By their honest toil, to cheer this visual gloom,
Fair Charity with kind, unwearied hand,
Supports the cause of virtue's chosen band.

Their aim is blessings on the blind to pour,
Make useful that, which useless was before;
Yes, charity will flow the useful grain;
And cheerful, industry each good obtain.

[Poem reproduced from the *Report of the Bristol Asylum,* 1799. No Author's name is provided. The poem is simply signed Philanthropist. From the Bristol Royal Society for the Blind]

A similar disdain of indolence and a belief in its cure by hard work and learning the Bible orally was developed by the founders of the Liverpool Asylum. Furthermore, correspondence between the founders of the asylum showed sympathy and pity for the plight of what they felt was the *poor blind.* This opinion was not just the result of uninformed, inexperienced speculation by these founders, as some of them were blind themselves and these issues featured in discussions of the liberal debating club where they met. In one of these letters one of the late blind founders of the Liverpool Asylum, Edward Rushton, a civil libertarian and anti-slavery campaigner (Hunter, 2002), suggested that a moral cure for the plight of those who were blind would be to employ them in paid work, where they could train for a future industrial trade, and in which they might find some form of happiness. To this end, in the excerpt from the correspondence printed below, it is emphasized that there is no worse fate than to be unable to work, to be *indigent.*[5]

> How piteous then is blindness! The face of the country, with all its various beauties; the town, with its docks, piers, and stately edifices; the aspects of his friends, of his dearest relatives, of his partner and his prattling offspring, are all to him a blank—are all involved in a mass of thick black clouds, which no summer's heat can dissolve, nor wintry storms disperse. How deplorable then is such an existence! Even with a competence, how cheerless! But with indigence, how dreadful!
>
> [Letter by Edward Rushton, founder of the Liverpool Asylum, to his colleague John Christie, a surgeon in Liverpool, 1790]

The British historian E. P. Thompson (2001) feels that such a moral disdain of poverty did not appear to be restricted to Bristol and Liverpool; it was a general feature in England during this period. This was a time of social and industrial change and, as a result of civil unrest and the large movement of people from the countryside to the cities, the number of beggars was rapidly increasing in all areas of the country from the end of the eighteenth century. As a result, Thompson found that in this period the cult of the *blind beggar* emerged. Although still associated with criminals and prostitutes, the blind beggar rose above a lowly status and became not so much a figure of loathing but one of pity. He quotes a famous saying of the time, *A blind man can get a guide at any place, because they know he's sure to get something* (Hunter, 2002, p. 293).

Similarly Thompson continued that whether or not society was becoming overrun by sexual immorality and crime or not, there was a general perception that it was. In his review of historical literature and surveys of the working class at the end of the eighteenth century, he reported that prostitution and crime were rife. For instance, in government reports of the late eighteenth century he observed that as many as one in ten in London were supposed to be harlots and criminals. However, on closer inspection he found that the classification of prostitutes included single mothers and unmarried women living with single men.

Along with a notion of the power of commercial handwork to cure indolence in the late eighteenth century, there was a belief that immorality should be met with a rigid code of behavioral control in the asylums. Thus docility as well as an understanding of God was achieved by the recital of biblical texts and hymn singing during working hours. This ethic was similar in schools for working class, *able bodied* students promoted by middle class philanthropists in the same period (Lawson & Silver, 1973). However, the emphasis placed on work, the bible and morality were stronger in asylums, where the working hours were longer, and the moral cure more *necessary* for what the rhetoric of the time felt was the particular need for morality in the blind (Hayhoe, 1995a, 2003, 2005a, 2006).

In a further illustration of this point, the register of the Bristol Asylum, which was first recorded in 1793, showed that sexual behavior was strongly discouraged in the early asylums, even between those of marrying age. The excerpt reproduced below shows that students had also been expelled from the asylum for conducting physical relationships, and as a result their potential livelihood from the trade that they practiced was denied them.

> [Jennifer], Aged 17 years a native of Ireland. Came in Dec 1796.
>
> Dismissed in January 1803. She acquired the knowledge of making fine [cane] work baskets, by which she could procure a comfortable livelihood, but having been guilty of very immoral conduct, she married [Jamie], one of the blind lads when they could no longer conceal their wickedness & both were dismissed, and are now begging in the City.
>
> [The Register of the Bristol Asylum, starting February 18th 1793. Found in the Bristol Royal Workshops for the Blind. From the Bristol Royal Society for the Blind]

By contrast, the same register also showed the rewards of hard work, diligence and docility in its training course. In this register it particularly likened the students of the asylum to its family, and emphasized a loyalty that was expected to this family. This feeling was often reciprocated by its students. There are particular instances in this register, for example, of students coming back to teach in the asylum or visit their friends and teachers. Other entries tell of the love of the asylum for the students, describe gifts of tools and money, or follow their progress to retirement and eventual death, as the excerpt below shows.

> [Sabrina] Aged 8 years a native of Bath. Came in Dec 1796.
>
> Left the 2nd April 1817 to make room for application of the admissions of others—was given 2 Guineas & a set of tools.

> [Venus] Aged 20 years, a native of Downend, came in March 1797.
>
> Died Dec 7th 1812. She was at the asylum from a few years after it was first opened: by her diligence & industry, she made a considerable proficiency in fine Basket Work, by which she procured a comfortable maintenance—She seemed studiously to promote Interest of the Institution & was amiable in her manners by which she obtained the love and esteem of the family.
>
> [The Register of the Bristol Asylum, starting February 18 1793. Found in the Bristol Royal Workshops for the Blind. From the Bristol Royal Society for the Blind]

This chapter now continues by examining the progress of handcraft usage in the nineteenth century, and compares it with its development at the introduction of the asylums.

The Use of Handcrafts in the Early Nineteenth Century

Evidence from this period of the century shows a continuance, even a firming, of moral attitudes and their effect on the curriculum of English asylums. For instance, the annual report of the Bristol Asylum (Bristol Asylum, 1838) in 1838 notes how its students, or what it referred to as its inmates, had become *happy* and *docile* through hard working days of weaving and what it refers to now as the mechanical arts. As a result, it felt that its students' minds were becoming more cultivated in the way that a sighted person gained knowledge from works of art or literature.

Similarly, in a survey of institutions for the blind published in 1833 the American educationalist and phrenologist[6] Samuel Grindley Howe (1833a), concluded that while the French system focused too much on literature at the expense of providing a trade, the English system too often overemphasized training for a trade and conducted excessive working hours of training. This, he felt, came at the expense of any form of literature or academic education.[7] As a result, Howe stated that he preferred the Scottish model, as it provided a more even balance of the observance of biblical texts and of trade.

Despite earlier ignoring it, the Reverend William Taylor, a vicar from Yorkshire, promoted touch literacy (Taylor, 1828) in the late 1820s and introduced it into an institution he founded in Yorkshire in 1835 (Warnock et al., 1978). Touch literacy remained a regional oddity for several decades afterward. Not until the 1860s was this introduced on a national level through two sources. The first what was an organization which was to become the world's first of its kind, the British and Foreign Society for Improving the Embossed Literature of the Blind, which was later to become the National Institute for the Blind, and is now the Royal National Institute for the Blind

(RNIB) (Thomas, 1936; Wolf, 1992). The founder of this Society, the English physician Thomas Rhodes Armitage who himself became blind later in life, was not influenced by the home-grown Yorkshire institution for the blind. His experience of literacy had been derived from his time on the continent, where he reputedly visited both the Viennese (founded in 1805) and Paris institutes during his travels and his medical education in Austria (Wolf, 1992). His reputation was built through influencing schools in England, and introducing Braille touch language in particular to many schools throughout the country.

The decision to use this touch language rather than one of the home-grown languages—the first such touch language was created by the English mathematics professor Nicholas Saunderson in the eighteenth century—was taken not only because Armitage experienced this form of literature on the Continent; it was also chosen because French literacy for the blind was much further advanced, as a result of its emphasis in the French institutes. Therefore, it made more sense to use this more evolved language and its printing methods. Nevertheless, he still felt society should guard against the indolence of people who were blind, and literacy should only be used as a means to enhance the vocational training that students were already receiving, which is why he stated *Training and employment [are priorities]* (Wolf, 1992, p. 4). Thus, despite the introduction of an element of Braille literature into the curriculum of the nineteenth century, almost all schools and asylums still placed an emphasis on handcrafts, while literacy was seen as an adjunct to the curriculum.

The second nationally introduced source of touch literacy came from a second direction, one that was unique to England in this century. Worcester College for the Blind Sons of Gentlemen was an elite independent school, founded for the education of blind children from the professional middle classes. It was established in 1867 by an Anglican minister, the Reverend Blair, who was also a master at Worcester's Kings' School (Craze, 1972), where he had taught the blind son of a friend and saw the need for a school specializing in this form of education. Several years after its foundation, it also collaborated with Armitage's institution to improve its provision of literacy. This link was to continue throughout the school's history.

A similar class differentiation was present in *mainstream* charity run schools for *the poor* and middle and upper class professionals in this period (Lawson & Silver, 1973). However, this class distinction was not as great as it was between those in this academic, middle class school and the more traditional asylums. In particular, the asylums had longer working hours and a more intense, boarding vocational curriculum than any schools for the *able bodied.* There is also no evidence to suggest that schools for the poor were wholly maintained by sales of goods as schools for the blind were. Therefore, although class was of a particular issue among students in general, it was a *particular* issue to those who were blind in the nineteenth century.

At this time, Worcester College was the only school to seriously consider a mainstream academic education in almost every academic subject as the aim for all of its students on entry, and admitted sighted students as well as those who were blind. This ethos continued until 1936, when its administration was taken over by the RNIB and it became known as New College, Worcester (Glanville-Smith et al., 1982; Wolf, 1992). Despite this change of administration, the legacy of Worcester College's independent foundation continues to exist at New College, as it still sends its principal to the HMC. However, similar to other charitable schools developed in this era, visual arts did not enter its curriculum until after 1981, with only music and literature favored as forms of creative activity before this chapter in its history (Fletcher, 1984).

This chapter now continues by examining further and more immediate financial incentives that handcrafts brought to their asylums and their later development.

THE ROLE OF FUNDING IN THE EARLY YEARS OF INSTITUTIONS FOR THE BLIND

In the early English asylums there was an emphasis on handcraft tasks for the purpose of raising money to fund asylums. This was also true of their Scottish counterparts. Early reports of the asylums in Bristol and Edinburgh feature the accounts from sales of goods from the students that supplemented the small donations the communities of the students were asked to provide.[8] These accounts in particular listed the individual craft products that were manufactured, their cost of materials and the net profit derived from their sale.

Similar attitudes were also commented upon by an educator and administrator in the Paris Institute named Guille (1819) in the early nineteenth century. He stated that the example of the English asylums in maintaining their buildings through this additional source of income should be employed by his institution, which was struggling to raise funds to stay in existence during this period. In this matter Guille (Paulson, 1987) had earlier disagreed with the founder of the institute, Hauy, from an early period and also felt that it should focus more on vocational training for the sake of its students' futures. This factor was particularly important to the Institute after the French revolution, as it was a favored charity of the king, and Guille, unlike Hauy, was a keen follower of the new republic. However, because of their cultural emphasis on a broader moral enlightenment, this aim failed and led to the closure of the institution.

This section continues by exploring the specific methods of fundraising and how they affected the curriculum of these institutions.

The Financial Influences in the Late Nineteenth and Early Twentieth Centuries

The pursuit of profit was not only a characteristic of the early asylums; it was also evident in end of year reports of the institutions in England throughout the later years of the nineteenth century (Bristol School for the Blind, 1887; Liverpool School for the Blind, 1849). Furthermore the *factories for the blind*, which evolved from the regional asylums and later local societies, were founded in the early decades of the nineteenth century to provide employment for students after they had left their schools and asylums. This ensured a continued income and a profit for the local societies that ran them (le Cue, 1992). Such factories still existed in England until the last decade of the twentieth century.[9]

Tables of accounts in annual reports from the nineteenth century also show that not only were students given vocational training, their success was also defined by the money they raised as a result of these crafts for their schools. For instance in the report of the Liverpool School from 1849, each student is listed in terms of their trade. Nowhere in the report, however, are their other educational accomplishments mentioned. In this report it is also noticeable that the students are found in the school *employed* at their trade rather than studying. This further emphasized the money generating qualities of their work. This is marked in the excerpt from the report of Liverpool Asylum shown below.

> [Sami], of the Parish of Sutton, Cheshire, aged 22 years, is boarded and lodged in the school, and employed at basket making. . . .
>
> [Jo], of the Parish of Toxteth Park, Lancashire, aged 12 years, is boarded and lodged at the school, and employed at sewing. . . . (Liverpool School for the Blind, 1849, p. 7)

The focus on vocational and literary education in England continued in many schools for the blind into the twentieth century, and that on the sale of goods was a strong motivational influence throughout the early part of the century too. For instance, a report from the Bristol School of Industry for the Blind (Bristol School of Industry for the Blind, 1908)—formerly the Bristol Asylum—in the first decade of the twentieth century described one of what it referred to as its regular exhibitions and sales of its work. The sale of crafts, it stated, played a role in supporting the Bristol School. This in turn gave extra impetus to the vocational, cognitive training of crafts. Such a commercial self-sufficiency was further reinforced by no mention in the exhibition notes of the aesthetic qualities or the gratification gained from the students' craft production.

The list of crafts on sale in the Bristolian catalogue, all created by the school's students, also reflected the crafts being taught in the Bristol School at the time. These appear to be dominated by those which were introduced at its inception, such as chair caning, and mat and basket weaving. The British educational historian W. H. Illingworth (1910) also reflects that many similar sales of work in schools for the blind were held throughout England in this period. For instance, in the appendices of his seminal work The Education of the Blind, he features many adverts from schools advertising similar sales of wares.

Correspondence to the Bristol school from the United States also illustrates the influence of the emphasis on handcrafts in English schools for the blind, almost 150 years after their inception. For instance, this emphasis on training and acclimatization to long working hours at the expense of literature appears to have been witnessed first hand by the director of the Perkins Institute, formerly the Boston Asylum, in 1930. In his letter—reprinted in part in the notes[10]—to the head of the Bristol School of Industry for the Blind, the principal indicated a close relationship between the two schools. However, although he expressed admiration for the head's skill in educating his students, and despite the Bristol School having an advanced academic curriculum at this point, he felt the handcrafts were still of too great an importance in comparison to his own curriculum.

In this letter, Perkins's director also insisted that he had no intention of prioritizing vocational work over academic study, as the Bristol school still did. However he did express admiration and asked for advice on the best methods of motivating students in order to emulate their long working days, as he was particularly impressed with the work rate of Bristolian students. Production was not only admired by the Bostonian in terms of the students' well being and their economic future; he also envied the school's means of financing their work via its factory for adults, and a society for the welfare of people who were blind in the community,[11] and so after this period a different use of handcrafts still survived.

The Development of Non-vocational Handcrafts

Signs that the nature of handwork in English institutes for the blind began to change during the 1930s exist in training literature written at the time. The RNIB and the College of Teachers for the Blind, a national organization formed to provide guidance on model practice in education, for instance, showed signs of looking at handcrafts not only for training in vocational tasks and economic subsistence, but also to develop their students' cognitive and tactile abilities. In particular, in their report from 1936, the Joint Committee of the College of Teachers of the Blind (1936) (JCCTB) initially referred to handcrafts more as a subject of tactile development for *practical*, noncommercial purposes (such as mobility and personal grooming) than an industrial craft. Furthermore, in its text the authors of the report also suggested that this new form of craft instruction should work alongside the past century's emphasis on vocational education, and began to make blind students' craft work more influential on, and reflective of, their personal and social lives as well as their intended vocational careers.

This report in particular emphasized a more holistic use of handcrafts for developing the use of touch for every day and educational purposes as, the report also finds, such tasks were not in widespread existence in the various curricula of schools for the blind at the time. Schools throughout England, it observed, still appeared to be placing as great an emphasis on vocational handwork as other elements of students' education.

In the 1950s this notion evolved further, and the JCCTB placed even more emphasis on the notion of handwork as an educational and cognitive tool as well as for economic subsistence. In a further report by the Committee of the College of Teachers for the Blind in 1956, a large section is again devoted to the uses of handcraft. Similar to the 1930s survey by the committee, this report also reemphasized the use of this craft form for enhancing practical abilities. Unlike the previous report, however, the authors suggested that the perceptual impressions of a practical form of handcraft, now called handwork, could enhance the emotional and psychological well-being of children who were blind, and benefit the mental ability of students pursuing their exercises for purely cognitive development.[12] Furthermore, in this report they went farther by suggesting that the practice of students using their hands to maintain themselves in every day practical activities, such as domestic chores, would provide them with more social freedom and a better appreciation of beauty through craftsmanship.[13]

> Standing as it does in a close relationship to art, handwork provides many opportunities to the blind child to express his or her own ideas in concrete form. Blind children should be trained from the earliest to appreciate beauty of form, proportion, craftsmanship thru handwork, and also realise that an article which fulfils its purpose, is well made, and well proportioned, is in itself a thing of beauty. (Committee of the College of Teachers for the Blind, 1956,p. 60)

This later form of craft work appeared to be the one in which students who attended schools for the blind prior to the 1980s were trained. In previous research, for example, I interviewed students who had been through these schools in this era (Hayhoe, 1995a,c, 2000). These suggested that basket weaving in particular was still the staple craft in these schools; however, no commercial use was made of it, as it only appeared to be used to promote docility in the students. This chapter now finishes with a summary of its findings in the segment below.

CONCLUSION

This chapter has shown that the curriculum developed in England was unique and had been affected to a large extent by cultural considerations of charity, the belief in morality as part of a vocational education and the financial welfare of the asylums' administration. These are now discussed under the following two sub headings.

The Marriage of a Commercial Culture and a Moral Disdain of Indolence

The original literature of the institutions in continental Europe and Britain all showed a fear and loathing of begging and the noncapital productivity it engendered. This factor was perceived to be a frequent source of income for many of the working and underclass people who were blind, and affected their moral standpoint. This belief was felt powerfully by the middle class, professional section of these societies and clerics, who formed the majority of the founders of the asylums in England. As a result, a moral solution was found for what was intrinsically a practical problem, and the first institutions were built to rid the streets of those in the worst forms of poverty, and to institutionalize and reform beggars whose presence was felt to be distasteful. It is therefore also arguable that the original institutions in Western Europe and Britain were formed on an ideal of moral social control.

However, although the institutions in France, Scotland, and England appeared to share the same disdain of indolence and the social implications of begging, the report by Howe indicated that they did not share the same solution to this problem. In this context, it appeared that the English curriculum was *radically* different. In particular, the commercial culture of English asylums reflected the culture of the Protestant industrial revolution, which ran through the eighteenth and nineteenth centuries. Rather than the study of literature, the English students' long hours represented not only training to raise them out of their perceived indolence and poverty, but a belief that such tasks would enrich their souls. In this belief, they were similar to other charity schools, although the culture in these asylums concentrated more on hard workloads and alleviating the *worst* affects of begging.

This belief was accompanied by student recitals of biblical texts and a strict code of behavioral control. This not only furnished them with greater morals, it also allowed them to remain at their commercial work throughout the day, made them docile and did not relieve them of their training tasks. In the words of the poem from Bristol, the task of the early asylum was to make *useful that which was useless before.* Besides, English asylums found that putting their students to work could fund the buildings and resources of these institutions, which received no government grants. Therefore, the economic imperative for the students to survive financially as individuals was not the only motive of these early asylums.

This chapter now continues by summarizing the final influence on the asylums.

Vocational Crafts and the Class System

The attitude of the English asylums in the eighteenth and nineteenth centuries, and the tasks they focused on as a result, also represent an amount of discrimination toward the lower and working classes. Despite being blind themselves, many of the founders of the asylums and later the RNIB were themselves middle class professionals and did not feel that their behavior equated to the indolence they reported in their correspondence.

Furthermore, the achievements of Saunderson, the Cambridge professor born blind, in much of the early literature belied a deeper scepticism of what could be achieved through their institutions. It is particularly noticeable in their early curricula that Sanderson's techniques of learning geometry or notation language were not employed in these asylums. It is also apparent in the nineteenth century that the only school founded on the notion of educating students who were blind from the middle classes did not entertain the notion of training its students in a manual trade. Therefore, although the potential for students who were blind was observable in this school, in *more charitable* institutions for the blind the students' abilities in these fields were still doubted.

Finally, although the emphasis on commercial handcrafts between the 1930s and the 1981 Act diminished, a belief in the positive nature of craft tasks to substitute an intellectual curriculum in England still reflected the earlier beliefs if not the skills of the earlier asylums. In particular, students were given the tasks of cane weaving baskets, a repetitive mechanical exercise which provided little creativity.

Its imperative at this time could also be seen as maintaining a skill that it was perceived would serve some practical use in the outside world, rather than provide intellectual stimulus as a solely aesthetic art. Furthermore, an element of commercial thought still went into the notion of cane weaving during this period, and many of the local societies that evolved from the original schools had factories which employed former students in order to maintain their usefulness, and used many of these techniques.

In the following chapter this book investigates the use of touch to develop intellect in the education of the blind.

NOTES

1. In his Letter on the Blind Diderot argued that women who were blind were in actuality more beautiful and moral innocent because they could not cast their eyes on immorality as women with sight could. In his Letter on the Blind it could also be argued that Diderot, a political radical who was imprisoned for his views on the French state and atheism, felt that touch was indeed a more honest sense than sight as tactile illusions were impossible.
2. Hence their moniker, camp followers.
3. Unfortunately, no source of accurate data collection about this disease appeared to exist during this period. However, as stated before tertiary sources such as Scholl and Stanier appear to agree about its significance as a cause of blindness and mental illness.
4. Circumstantial evidence also appears to suggest that syphilis could have been a major problem in the male homosexual community at the time, leading to a belief that they were susceptible to blindness more than other. In recent decades, for instance, data from the medical records of homosexuals have been analyzed separately in epidemiological studies, and have shown that male homosexuals are more likely to contract this than male heterosexuals (Stanier et al, 1987).

   Michel Foucault (1989) in *Madness and Civilisation* argues that this homosexuality was not only a moral affront to European society in the eighteenth century, but also classified as a form of mental abnormality, or illness. He further argues that those who were homosexual were felt to be intellectually abnormal. Porter (1990) also finds evidence in the eighteenth century of the bitterness shown to male homosexuals, arguing that in the eighteenth century mobs attacked these men. Although these attacks were relatively rare, Porter suggests, they appear to reflect a negative attitude toward homosexuality during this period.
5. It should be noted, however, that the author of this letter was blind himself, although from a wealthy, middle class family and held a post in a company in the city after becoming blind.
6. Phrenology was a popular science in the nineteenth century, and was a forerunner to psychology. Phrenologists believed that a person's personality and intelligence could be gauged by the physical size of certain parts of their skull, which indicated the size and shape of the brain underneath the correspond-

ing part of the skull. Phrenologists believed that size of brain was inherited and also changed over time dependent on exercise and stimuli.

7. Howe was examining European models of educating students who were blind in order to form a model for a similar asylum in Boston—later to become the Perkins Institute.
8. See Le Cue (1992), RBAS (1993), Bristol Asylum (1800) and other instances given in Hayhoe (1995a,b).
9. After reading a paper I presented as a lecture (Hayhoe, 2005b), I was sent an email by a teacher in a school for the blind in Kansas, US. He found that factories for the blind still exist in his state. The following is an excerpt from his email.

   > There is an organization here in Wichita that employs hundreds of blind people at tasks that were replaced by automation decades ago. I was much taken aback when I visited one of their departments and saw a room full of people assembling four plastic parts to make 29 cent ball point pens. The broom factories of the past were more fulfilling of a sense of self worth. The experience gave a stark realization of our focus on what they don't have rather than what they do have.

10. Dear Mr. Stevens,
    My grandson, David Malan, of Old Malthouse School, is here visiting us. On cold days he wears his grey sweater made by you; and we like it as well as he does. Though I know you make these articles, I don't know how you sell them. Do enlighten me on this process.

    Dr. Ritchie says in his excellent book that we American blind school men conduct our works departments half-heartedly. He isn't far wrong. As for Perkins it has tried hard to have its corresponding department of manual training, educational and prevocational rather than vocational. Now since in this country you never can tell what a fellow will follow in life, our plan has worked very well indeed, so long as pupil material remained largely promising and the times kept propitious; but now that these two elements are less favorable, we feel called upon to experiment in industry and need advice how best to do it successfully at school. . . .

    My school is a school and maintains the atmosphere of one. Everybody works, yes; but the motive power is desire through leadership. Compulsion isn't in evidence. Doubtless we carry this spirit to an extreme. When the recess bell rings, as it does frequently, everybody responds very readily, and pretty much everybody looks upon his extra-curricular activities as decidedly important. American undergraduate youth does everywhere. When a legal holiday arrives they claim it as their right. How in an atmosphere of this sort am I to introduce my industrial conditions?

    The British blind school manages to maintain such conditions. How does it do it? Your big school boys learning basket making dress for work, sit on the floor and labor hammer and tongs as though life—or dinner—depended upon it. How long at a time do they keep this up? How many hours a day? a week? What is their incentive? Is it pay? Piece pay or saleable product? And your large girl apprentices still at school—how do you motivate their anxiety to learn and produce as evidenced in their diligence?

Your Mr. Gray showed me his new means of keeping his big boy apprentices apart from the rest, except at meals—a separate and distinct rooming residence. Now I like his plan and approved of the idea though I did not mean to follow it here. My present plan is either to send regularly a small voluntary group of my older boy apprentices away from the school premises for several consecutive hours of intensive labor at an industry, but to have them return as regularly for some minor activities, academic and social, and of course for living, betweenwhiles; <u>or</u> to try to carry out all these activities, major and minor, here at the institution. You people do it; perhaps I can. That is to say,—when Dr. Ritchie claims that we press our works departments only half-heartedly, the natural inference is that he and Mr. Gray and other headmasters do it whole-heartedly. I seek the secret to apply to you to acquaint me with it.

Bristol has become a pleasant memory to me. You enabled me to see and understand more both of city, shop, hostel and home life than has been customary in my peregrinations, and I thank you heartily. I mean to run over again some day. Meanwhile shall you not visit over here, perhaps with other Europeans at the international congress next spring? I should love to entertain you.

Cordially yours,
Edward E Allen

[Letter to the Head of the Royal Blind Asylum Workshops (formerly the Bristol Asylum) from the Director of the Perkins Institution and Massachusetts School for the Blind, United States, dated September 8, 1930. From the Bristol Royal Society for the Blind]

11. Handcrafts produced in a factory for the blind existed as a source of income in the Bristol Society until the 1990s, though the school is now closed.
12. It does not appear to be clear from the evidence from interceding reports how this belief changed. It is recommended, therefore, that more research appears to be needed into this subject.
13. Viktor Lowenfeld proposed similar changes in the practice of mechanical arts in U.S. institutions in the same period. These he argued were continuing to be emphasized at the expense of more intellectual or aesthetic tasks using hands as a form of direct perception. This was caused by the understanding that art was of little use unless practiced by more elite members of society.

    [Art] for the blind was considered misleading the blind, and diverting them from the preparation for their vocations which they needed to earn a living. That they can earn it better as well adjusted individuals—and that creative activity used as an emotional outlet helps them in their adjustments—has not even penetrated the thick walls of most of our institutions today. (Lowenfeld, 1951, p. 1)

CHAPTER 5

AESTHETICS AND TOUCH IN ENGLISH SCHOOLS FOR THE BLIND

INTRODUCTION

Previously this book described how the evolution of cultural knowledge about blindness from the age of scientific Enlightenment changed for economic and theological reasons, and the attitudes of educational theorists in England were changed according to religious morality and economic reasons. Whereas before blindness had been linked with immorality and even evil, after it people who were blind were thought to be able to be changed to become morally useful, and have perceptions that could be thought of as roughly approximate to, although not as powerful as, sight.

This chapter discusses how the attitudes of these institutions affected the creation of aesthetic education in English asylums and schools for the blind. Furthermore, this chapter also shows that despite their use of scientific arguments, these institutions reflected the cultural, political and religious cultures in England more than they did the thoughts of Locke, Molyneux, Newton, Saunderson, and Berkeley.

God, Money, and Politics, pages 77–91

MUSIC EDUCATION IN ASYLUMS AND SCHOOLS FOR THE BLIND

In the first English asylums, music was felt to be the aesthetic form that people who were blind could appreciate more than others (Bristol Asylum, 1800). The influential papers on education by Demodocus (1774) and Hauy (1889) in the late eighteenth century had both recommended music as a talent unparalleled in people who were blind, suggesting that in this subject people who were blind even had an advantage over people with sight. This opinion was justified, they felt, because they had observed the skill of the beggars who were blind busking instruments for a living, and appeared superior in their ability.

These authors and the asylums' founders also stressed that these musical abilities could be put to greater use in more moral arts, such as church music or the mechanical preparation of instruments. Like the use of handcrafts, therefore, the earliest asylums' curricula echoed their vocational training and moral, philanthropic foundations. These views were not new, however. The same observation was made earlier in the eighteenth century by the French philosopher Denis Diderot (2001) in his *Letter on the Blind for Those Who Can See*, which like the theories of Locke, Berkeley and Saunderson, was mentioned in Demodocus and Hauy's essays.

In the composition of his letter, however, Diderot had motives that were very different from those of the later *educators of the blind*. He used the skill of blind beggars to illustrate the mechanical and aesthetic ability that blind people could achieve, rather than its Godly merits or vocational attributes. Although addressing the philosophical concept of aesthetics and reproduction, his radical symbolism of the ability of the blind and the nonvisual senses flew in the face of conventional, particularly religious wisdom.

Diderot's symbolic use of blindness was similar to that of his English counterpart, Locke, a century before, and he was a writer that Diderot greatly admired. Like Locke, he had also written a controversial work, *The Encyclopaedia of Philosophy*. They were also at one point political dissidents. He had been denounced by those in authority, which had led to his imprisonment by the King of France, much like Locke's association with radical Protestants, which antagonized the English monarchy's soldiers into hunting for him.

Both also descended from positions of great political power. Diderot had previously held positions of political influence with the King, much in the same way that Locke had done in the parliament and economy of England. Therefore, the manipulation of music with the aim of manipulating moral reasoning flew in the face of such radical depictions, and provided a symbol of anti-establishmentarianism. This process of moralization of this art is now discussed below.

Notions of Musical Aesthetics in the Curricula

The earliest English asylums' curricula only permitted the use of church music as a form of Christian worship, allowing the inmates to continue their handcraft work unhindered while they practiced it. This also allowed them to rehearse pieces for their regular Christian worship (Bristol Asylum, 1800). Likewise, music served two further fundamental purposes during their education: it was a form of art which had a substantial nonvisual moral element, and it was a way of increasing workers' docility while they worked. In addition, there is evidence that the asylums felt that they could take the use of this subject farther, and change it into yet another trade that could elevate their inmates from indolence. This changed the practice of music to tasks that would become less emotional, such as rote learning of standard pieces, and marginalized its use of aesthetic expression.

The English experience of music education differed markedly from that on the Continent in the latter part of the eighteenth century and the first years of the nineteenth century. These institutions felt that music was an art in which blind people could gain confidence and prove an ability they felt was lacking in other mental activities. However, rather than proposing a more liberal use of music by most of their English counterparts, their theory on the use of music was one of emphasizing what they felt was the deficit of their other faculties. This point was highlighted in the early years of the Paris Institute by Hauy (1889), who expounded what he felt originally were mental defects in people who were blind.

By the 1830s, a number of the English institutions began changing their attitudes to music to more closely reflect a belief that it could be taught as a subject in its own right. For instance, during this decade the Bristol Asylum introduced a professional music teacher to carry out separate lessons. Unlike Hauy, though, he reflected their Protestant belief that they had discovered that music was a trade they could excel in, to provide its students with further opportunity of employment at the end of their education. This is illustrated in the excerpt below.

> Practical utility has ever been a principle before the committee; thus the musical study of the pupils has always been directed to the training of organists; and at present, [Henry H], one of their number, is an assistant organist in the chapel. . . . (Bristol Asylum, 1838, p. 6)

Similarly, in the same decade US educators concurred; although teachers such as Howe (1833b, 1836a,b, 1837) were in agreement with Hauy in the early years of his asylum (Paulson, 1987) in believing that, apart from a limited number of oral and aural skills, students who were blind would never be the equal of those who were sighted in such academic or mechanical subjects as this. In fact, in 1837 Howe went as far as to say in a lecture on phrenology, disability and race—a topic he studied assiduously—that the brains of the blind had changed to weaken their understanding of many mental concepts. This, he felt, supported his argument for the introduction of eugenics, which could prevent people with perceptual disabilities from having children.[1]

This chapter continues by examining the influence of music on the curriculum of English institutions in the latter half of the nineteenth century.

Music Training in the Late Nineteenth Century

With the introduction of The Normal College and Academy of Music, formed by Thomas Rhodes Armitage among others, and Worcester College for the Blind Sons of Gentlemen, the late 1800s emphasized music education not only in curricula but also in the full title of the Normal College. This underlined the belief brought about by Armitage that students who were blind could study this art with equivalence to their sighted counterparts (Thomas, 1936; Wolf, 1992).

There were two further marked differences in the emphasis of music in the curriculum in this era. The first was its effect on the quality of musical education. For instance, as its responsibility was to improve the value and educational circumstances of people who were blind, the Normal College undertook more structured music lessons than previous asylums attempted to, and provided extra time for this element of the curriculum (Thomas, 1936; Wolf, 1992). The new curriculum of the Normal College became so successful that it had an effect on other institutions for the blind in England (Illingworth, 1910), and encouraged them to provide extra time for music lessons and preparations for concerts. True to their original economic aims, however, evidence also shows how this extra time was turned, like handcrafts, to raising funds for their institutions. For example a report from the Bristol School for the Blind (1887) describes the outcome of this extra tuition in terms of increased attendance at its monthly fundraising concerts.

At Worcester College, music was also given an elevated place in the curriculum, and their aim was for a far superior education that would allow their students to enter the upper echelons of the outside world (Worcester College, 1879). These were funded in the main by private bequests, either from relatives of students, local philanthropists or former students. The rhetoric of these bequests to Worcester College also reinforced a belief in the particular advantage of the college's students pursuing a musical education to provide a future profession (Worcester College, 1881). Ironically, despite its more middle class foundation, this notion was similar to the Protestant motivations of the asylums, and demonstrated how the incentive of skill and vocational expedience, despite the different levels of attainment, also influenced this choice of art as a symbol for conformity in the English culture. The extract below is an example of this.

> [As] far as we understand the matter, [the bequestor is] . . . strongly of the opinion that a musical education is the best possible training that a blind boy can have; that success in music . . . is more likely than in any other branch of education; that therefore more self supporting persons are likely to be made by this art than any other. (Worcester College, 1881, p. 14)

The difference in the level of music curricula between Worcester College and other institutions for working class, poor children who were blind is highlighted in the higher academic expectations of children from middle class backgrounds at Worcester College (1881; Thomas 1936). In this respect, the belief in the weakened ability of the blind community extended only as far as its working class elements. Not only did Edward Elgar and other Worcester Cathedral organists count among the teachers, the students also learnt academic, and by extension, the aesthetic elements of music, which also became a symbol of their social superiority.

The music lessons in the Bristol School (Bristol School for the Blind, 1887) at this time, for instance, were still very simple, concentrating on the memorizing of hymnals and piano pieces as schools such as the Normal College (Hayhoe, 1995a; Illingworth, 1910; Thomas, 1936; Wolf, 1992), were only training local organists, choristers, or manual music professionals, such as piano tuners. By contrast, the extract from Worcester College's annual report exemplified their more rigorous approach.

> On questioning the boys on the theory of music, I found them thoroughly well acquainted with the nature of clefs, scales, time, signatures, etc. . . . The system carried out by Mr. Done [Worcester Cathedral's organist] is calculated to impress his pupils with a genuine love and appreciation of music as an art, and thus encourage their efforts to make real progress year by year. (Worcester College 1881, pp. 21–22)

The students from Worcester College also received assessments from external academics, which other schools did not (Worcester College, 1881). This conformed to the view that students at the college should be prepared for university entry, and integrate with what were regarded as their academic peers despite their blindness, again a sign of their expected preeminence in the greater community. Despite their encouragement in music and the increased belief within Worcester College about the intellectual musical ability of its students, evidence from the latter years of the nineteenth century showed that this attitude was not shared by the most prestigious institutions of *mainstream* society (Hayhoe, 1995a,b). Although the records show that Oxford and Cambridge colleges, after initial reluctance, accepted students from Worcester College and often provided scholarships to the most proficient, other elite universities and colleges of music felt differently.

For these institutions the leap of faith from their orthodox standpoint seemed too large a gamble to take, and too large a bastion of intellectual faith to concede. Their most needed beliefs were challenged by these students who were achieving what stood against the most strongly held tenets about educational achievement and physical ability in this period. For instance, a report of Worcester College from 1889 reported that students at the college had been excluded from the public examinations and auditions for the Royal College of Music and the British Academy, as it was believed that they were incapable of *normal* study. Similar sentiments were articulated about students who trained at the Normal College, although this ability was questioned by less lofty organizations. In the same year (Salmon, 1889). it was reported that the College had also found difficulty in having any of its students considered for posts as church organists, as employers did not believe it was possible for people who were blind to play such pieces. This attitude would later crumble, however, as the social attitudes toward liberalism and a growing acceptance of blindness in not only music but also other forms of art strengthened as England enter the twentieth century. This new emphasis on the long path to acceptance and creativity within music education in the twentieth century is now examined below.

Music Training in the Twentieth Century

In the first years of the twentieth century, the study of oral and aural music was still the aesthetic task favored over others in the curricula of English schools for the blind. It was also considered by Campbell (1901), one of the founders of the Normal College and now its head teacher at the beginning of the twentieth century, to be the most important subject that a student who was blind could study. However, despite his affirmation of the importance of music, Campbell also felt that in the more charitable schools, including his own, whose population was mainly from working backgrounds, the teaching of music was still lacking. As a result he stipulated that for the subject to be fully appreciated it must be learnt through the compositions of the best composers, through concerts and recitals.

It also had to be taught, he continued, more from the perspective of an academic subject. It should be appreciated as an art in order to be followed efficiently as a profession. Thus, Campbell felt that a similar system of musical appreciation such as only previously privileged to middle class students at Worcester College, should be available to the future piano tuners and church organists, taught at charity institutes for *the poor*. Below is an example of this view.

> If the mental faculties have not been developed and thoroughly disciplined, the blind music teacher or organist, however well they may play or sing, will generally be a failure . . . music instruction in its several branches of harmony, piano forte, organ, and vocal culture should be addressed to the mind, not merely the ear. (Campbell, 1901, p. 7)

Despite this early belief in the superiority of music as an aesthetic art form, a later report by Thomas Armitage's National Institute for the Blind (Joint Committee of the College of Teachers for the Blind, 1936) in the 1930s showed that Campbell's earlier aspirations had not been accepted by their other schools, including Worcester College, which had now fallen under the auspices of its administration and welcomed students from a broad range of social backgrounds. Although it was also argued that music was the art by which students who were blind could derive aesthetic pleasure, the report observed that music was still considered an *ornament* of the schools, and a method of making a living in a more elementary sense after finishing school, rather than as an aesthetic art to promote cultural development.

In my research into students' experiences (Hayhoe, 1995a, 2000), I have made similar findings. During interviews with students who had attended schools for the blind from the 1950s to the mid 1970s in particular, the students told me that the oral and aural culture of music seemed to be particularly prevalent, with musical theory and notation having a lesser role. In particular, choral works were favored more than any other and even instrumental teaching also took a secondary role during their lessons. These findings were predominantly confirmed by the description of music teaching from the 1960s until the 1970s during the formation of the 1981 Act, at RNIB New College, Worcester.[2] The then head teacher, R. C. Fletcher (1984), later reported that simple performances at events, such as parties and nonmusically based local festivals rather than the academic study of music, targeted at university entry or professional performance, were encouraged at the school.

Despite these observations, before the 1981 Act, the aesthetic appreciation of music still reigned as the most influential aesthetic art in schools for the blind. However, in many other lessons direct experience and intellectual aesthetic appreciation were also integrated through haptic tasks, although they were rarely acknowledged as a form of art. These tasks are now investigated in the segment below.

AN INVESTIGATION OF AESTHETIC TASKS IN NON-AESTHETIC SUBJECTS

Evidence exists that the aesthetic appreciation of nature was part of education from the second and third quarter of the nineteenth century in England. For instance, the Rev. Taylor, who founded his pioneering institute in Yorkshire in the 1830s, believed that a touch education relating language to real world objects, such as that pursued by Professor Sanderson in the seventeenth and eighteenth centuries, provided enough information to sustain a full academic curriculum. Furthermore, in the 1850s the Bristol Asylum (Bristol Asylum, 1854) conducted a form of the aesthetic enjoyment of the beauty and humor of nature during occasional nature walks, and visits to local botanical and zoological gardens. This even led to students producing forms of the flowers and small animals they had felt with their hands. Nevertheless, more concrete evidence of tasks involving the production of artifacts in academic studies did not appear in England until the later decades of the nineteenth century at the Normal College. For instance, in his article in the late 1880s, the British journalist E. Salmon (1889) observed glass cases with clay models of flowers, seed pods and leaves formed by students in the Normal College academic classrooms, as part of their nature studies.

The Normal College's adoption of this task seems to have emerged from the observations of the earlier medical education of its principal founder, Thomas Armitage (Hayhoe, 1995a; Wolf, 1992). It should also be noted, however, that the use of handcrafts in making and understanding maps and globes is also evident at Perkins Institute. These were also developed from a long established relationship with the Vienna Institute for the Blind in this period, and the cofounder of the Normal College, Campbell, was a teacher in this school. During his studies in Vienna, Armitage was said to have spent a great deal of time visiting the Viennese Institute, which had been founded and run by the German educator and social reformer Johan Klein in 1805 (Klein, 1971). He was unique among the educators of the blind at the time, as he was the first to believe it was possible to provide a fine art education for students who were blind, even in those born and raised as blind students.

According to Geza Revesz's (1950) case studies of artists who were blind, Klein trained the first known totally blind artist, the Tyrolean crucifix sculptor Kleinhans, at the end of the eighteenth century. Klein also appears to be the first educator to propose the first non-vocational use of handcraft tasks for the emotional and cognitive development of students who were blind. Like the motivation of commercial handcrafts by English Protestant foundations, Klein's Catholic belief in the practice of this form of art appeared to be motivated by the use of craft and art to prevent immoral behavior. In particular, in his seminal paper on his theory of education, first published in German in 1836, Klein suggests that the use of hands to play with toys, mold and sculpt was particularly useful in the prevention of masturbation. This, he believed, was more of a problem in blind students, whose lack of visual stimulation would lead them to discover other sensual paths, particularly those related to unregulated touch. An example of this view is illustrated in this excerpt from his paper.

> [The] blind person who cannot be stimulated by vision and who is thus used to gaining pleasure from feeling objects is more liable than others to involve himself in the vice of masturbation that weakens the body and soul. Exercise and occupation can prevent this most effectively. (Klein, 1971, p. 235)

Therefore, in its nature classes the Normal College inherited two of Klein's educational theories. The first was his belief that blind students were capable of aesthetic tasks in a way that was comparable to students with sight. The second was that substituting natural objects to create emotionally intellectual and aesthetic stimulation would replace a different form of immoral emotional stimulation, fed by a notion that, left unenthused, human minds will tend toward sexual depravity.

Klein's theory that blind students were capable of equivalence to sighted students was not shared in Austria in the latter years of the nineteenth century, however. For instance, at the first International Congress for Teachers of the Blind in Vienna in the 1870s, attended by many educators from England including Campbell, the teacher M. Pablasek (1873) from the Vienna Institute stated that he doubted whether the use of handwork could be utilized for aesthetic or intellectual tasks with any effect. This belief also affected Austrian education for the blind into the twentieth century. For example, as this book discussed in chapter three, in the 1930s the Austrian art teacher Viktor Lowenfeld found that he was treated with disdain for suggesting he could teach sculpture to a girl who was deaf-blind in his early teaching posts (Lowenfeld, 1951; Lowenfeld & Munz, 1934). It was not until he left for America, and established an academic reputation in the US in the 1950s that he was able to put these measures into greater effect. However, even there he found it difficult to gain acceptance.

The development of this non-vocational arts and crafts education in the twentieth century prior to 1981 will now be examined below.

Aesthetic Tasks in Subjects During the Twentieth Century

The traditional role of handcrafts as a non-aesthetic art was challenged in England from the US in the early years of the twentieth century. For instance, a training manual from the Perkins Institute for the Blind (1906) that was used in English schools questioned the nature of the use of handwork tasks, and maintained that a greater emphasis should be placed on the use of handwork as a tool of direct, experiential learning. Moreover, the manual writers felt that handcrafts had the potential to be used as a means of self expression and information gathering in the academic curriculum.

As this book described in chapter three, unfettered by the culture of existing schools for the blind, the curator of the Sunderland Museum and Art Gallery James Charlton Deas (1913) also designed an unrivaled course for students from the local school for the blind visiting his institution in 1913. Tasks designed by Charlton Deas were based on a touch tour of his museum and were followed by a series of classroom exercises in the museum's public areas. Charlton Deas's tasks included the modeling and reproduction of artifacts available with clay in the museum. In his discussion paper on this course, he observed that students worked enthusiastically in their creative tasks in particular and appeared to show an understanding of the artifacts he was portraying through the models they produced, and through their descriptions of the artifacts. However, he also reflected that he was worried that the funding and appreciation of such initiatives appeared to be severely under threat in a period in which the British economy, at the height of its economic Empire, should be able to afford public institutions that could experiment with new forms of liberal arts. Instead, he found that budgets for such public institutions were being withdrawn within the year in order to pay for armaments for the impending war.

In a paper published in the latter half of the twentieth century, the British curator James Bartlett (1955) found that Charlton Deas's worries were founded, and that his early experiments had little impact in the country after this period. Although supported by a brief movement in the United States following his initiative, the showing of museums to students who were blind reverted to an ad hoc system of visits organized by teachers, with little direction or thought to the curriculum in the following decades. In particular, in these years few resources had been made available to visitors who had any form of disability. Charlton Deas' fears over lack of funding were particularly prevalent in the years immediately after World War I. After the armistice, more resources were concentrated on rehabilitation and the retraining for work of people who were blinded in the war (RBAS, 1993; Thomas, 1936). Similarly, few exhibitions or courses for students who were blind appeared to be organized in museums after World War II, and the whole initiative stagnated until the last quarter of the twentieth century, when disability rights became part of the broader civil rights movement a few years prior to the 1981 Act (Fergusson, 1987; Pearson, 1991).

Meanwhile, aesthetic hand tasks within nature studies, history, geography, mathematics and science continued in schools in England in the latter half of the twentieth century, much in the same way as those observed in the nineteenth century. In interviews I conducted with students educated in this era (Hayhoe, 1995a, 2000), it was observed that in isolated classes students continued to model natural objects and build landscapes in the latter half of the twentieth century.

In these intervening years, institutions continued to believe that students who were blind were unable to perform *real* fine arts tasks or appreciate works of art in more formal lessons, which suggests that they believed the sensations derived from tactile perception did not allow aesthetic equivalence to sight. Evidence that this notion was also true of a belief of the lack of intelligence and academic ability of these students as a whole also exists during these years. In particular, in his reflections of his school career the blind former British government cabinet minister, David Blunkett, described being excluded from mainstream examination courses (Pollard, 2004). Eventually, and against the wishes of his head teacher, Blunkett had to enrol on a night school course at a local college to gain many of his formal qualifications.

Evidence of intellectual experiences similar to these of Blunkett is also described in my own research (Hayhoe, 1995a, 2000, 2005a) and that of the British educational psychologist Sarah Dharma Colwell (1993). In studies which included oral history and observation with case studies from the latter years of the twentieth century, we have found that there was a distinct difference in the positive school expectations of students who were blinded later in life and the negative expectations placed on those who attended schools for the blind in early life. Furthermore, the views of a prominent educationalist, R. C. Fletcher (1984), serving as head teacher of New College in the 1960s to the 1970s, provided additional evidence of a belief in the social and cultural inferiority of such students in many intellectual contexts. Fletcher in particular appeared to believe that his students' ability to gain information from direct experience of tactile perceptions did not allow their necessary understanding of academic subjects, so that they could study them in as great a depth as a sighted person could. This lack of necessary information, Fletcher felt, left a grey area in his school's understanding of the aesthetic appreciation of his students in many of their subjects. They therefore could not, he argued, be expected to study with equivalence to those with sight. His only exception to this belief again appeared to be tasks relying on music as an oral and aural culture.

CONCLUSION

> What we have [in this country] is socialism for the rich and free enterprise for the poor. (Ghandi Memorial Lecture, at Howard University. Delivered on 6th November 1966 by Martin Luther King)

The evidence discussed in this chapter suggested further conclusions about the origins of exclusion from a fine arts education for the blind in England. In particular, this chapter presented evidence that the place of musical and craft-based aesthetics in institutions for the blind, both as a discrete subject and as an element of other academic subjects in England, was motivated by economics, moral considerations in the curriculum and a division of social classes. In these institutions there was also a belief that the oral and aural culture of the blind was superior to that of tactile perceptions. Furthermore, evidence presented in this chapter also showed that, prior to the 1981 Education Act, a complex curriculum was based on a social belief in schools for the blind that tactile information could not provide the depth of information required by students in aesthetic subjects, despite its use in literacy through Braille. These issues are discussed in more detail below.

The Social and Economic Emphasis on an Oral and Aural Aesthetic Culture

Despite seemingly successful educational experiments in Austria from the late eighteenth century, tactile fine arts were rejected in English institutions until the 1981 Act in sole favor of music. There were three main reasons that aural and oral aesthetic cultures were favored over these.

First, it was believed that music was a form of aesthetic art that could be enjoyed *in conjunction* with handcrafts, rather than extra time being made available in the curriculum for separate lessons. This was particularly noticeable in the early institutions, which chose choral tasks rather than training in the use of instruments, and favored oral and aural learning rather than music notation, which would interfere with their commercial work. As asylums chose not to include literary or academic curricula as institutions on the European Continent did, less time was taken from students' vocational training and commercial manufacturing. In later years, when musical instrumentation was introduced into these institutions, their teaching again focused on using this art form as a way of earning a living, rather than appreciating its art form. Most notably, church organists, piano teachers and tuners, and choristers were trained. In contrast, fine art or literature was seen in this period as unlikely to provide sustainable careers for working and under class students with visual impairments, as many of them were street beggars and the offspring of syphilitic whores and other immoral unions.

Only Worcester College, whose much smaller number of fee paying students came from professional middle class backgrounds, appeared to introduce an academic, high culture study of music. This amplified the class difference in the education of the blind in the last half of the nineteenth century and the twentieth century. It also served to reinforce the source of social and cultural attitudes shown toward the mental ability of students in English charitable institutions, and the roots of a curriculum designed to relieve poverty, indolence and unethical trades. In this context, blindness came to be regarded by those who wished to sustain their financial and social power in English culture as a symbol of liberal pity, and therefore their own superiority.

Second, the earliest asylums believed that their curriculum could increase the moral outlook of their students. Without sight, however, it was felt impossible to understand works of knowledge, art or nature first hand. The only other artistic method of acquiring beauty or intellect therefore was an oral and aural culture—in England and countries such as France and the United States, touch as an aesthetic tool was initially frowned upon for any other purpose than vocational crafts; although in Vienna it was overtly linked with sexuality by Klein. The worship of God through the beauty of choral music was, therefore, a significant element of the patronizing attitudes of the liberal establishment which promoted a sense of accessible moral nourishment. This was particularly evident in the choice of Christian choral music and in the training of organists for employment by parish churches. Apart from its use as a symbol of the cultural superiority of the philanthropic middle and upper classes, music also had a further purpose. It was used during craft tasks as a method of creating docility during the hours of labor, and was thought to stop the deviant behavior that blindness was associated with. The rhythms of the singing would also have aided the work, making the students more productive. This use of music would therefore have been compliant with the early institutions' philosophy that their curricula were designed to reform immoral beggars in particular, people who were especially indolent.

Third, this chapter showed that many prominent educators appeared to believe that students in charitable institutions were incapable of high levels of intellectual or aesthetic thought, such as those promoted by intellectual arts. The essay that Hauy used to gain funding and royal patronage, for instance, demonstrated the belief that blind people were lacking in these faculties. Similarly, in England the idea of including fine art or other intellectual subjects in the curriculum was thought to hold no overall benefits to the ones that could not be better served by vocational training. In schools run by charitable institutions, this belief was emphasized by their choice of employment. Piano tuning and repetitive organ recitals were favored over others that would have required greater levels of thought or creativity, such as music composition. It is particularly noticeable that the practice of piano tuning required little but the mechanical use of simple tools to achieve an unambiguous aim. The only judgement required in this task was to recognize a predefined note.

Aesthetic Tasks in Non-Aesthetic Subjects

This chapter presented evidence that the tactile appreciation of nature was first introduced in the nineteenth century. This perspicacity, however, was not linked to the plastic production of works of art until the introduction of an Austrian style, noncommercial handcraft in the latter half of the nineteenth century, with the founding of the Normal College. Rather than purely educational, however, the motivation for the use of modeling was again moral. For example, Klein, the founder of the Vienna Institute—an institute that particularly inspired the curriculum at the Normal College—emphasized the use of handcrafts to prevent immoral uses for untrained hands, such as masturbation. Unlike its earliest Austrian influences, the use of hands to produce models was still not considered an aesthetic form of creation in England. Instead, this task was linked to the notion that handwork to reproduce natural objects in particular was only designed to facilitate intellectual understanding. Additionally, doubts were expressed about the efficiency of touch compared to that of sight for providing sufficient intellectual information, and the ability of students who were blind working with hand crafts in charitable institutions were continually doubted.

By contrast, these beliefs were not applied to students in schools which catered for blind children from middle and upper class backgrounds. At Worcester College in the latter half of the nineteenth century, there was no doubt that their students could achieve high academic results. For that reason, these students continued to be educated in academic subjects to get places at top mainstream universities. Class, therefore, had a significant effect on the design of tasks in the curriculum of schools for the blind. It also influenced the beliefs of educators in the nineteenth century, who were unsure whether tactile perception could help intellectual thought or restrict it. This debate was continued into the twentieth century by educators such as Fletcher—the head of Worcester College after it changed its administration to the RNIB and became New College, and accepted students whose fees were paid by local authorities—who believed that his students' ability to understand through touch made their intellectual powers inferior, both intellectually and aesthetically.

In the following chapter we investigate the 1981 Education Act, and its evolution from a report on integration.

NOTES

1. In this lecture he made a similarly disparaging statement about people "from hotter countries," whom he claimed, although they should be treated with equality—he was against slavery—were intellectually less capable than those from colder climes.
2. Formerly Worcester College until it was taken over by the RNIB.

CHAPTER 6

LEGAL INTEGRATION IN ENGLISH EDUCATION

INTRODUCTION

This chapter examines the move toward legal integration of blind students in English mainstream schools, and more important the economic and political motives for planning the integration of disabled students beyond the more altruistic aims suggested by the government at the time. We present this description in the following two sections: the development of information gathering, and an investigation of the passing and initial implementation of the Act.

AN INVESTIGATION INTO THE DEVELOPMENT OF INFLUENCES AND INFORMATION GATHERING PRIOR TO THE 1981 ACT

An initial movement toward an understanding of the growing political movement promoting the social inclusion of people with disabilities resulted from discussions between members of the Conservative government and what is now considered an early form of the disability lobby in the early 1970s. To understand the formation of this lobby, however, we have to look back more than a hundred years to the history of British colonial history. Disabled people in Britain had become initially politicized after the return

God, Money, and Politics, pages 93–105

of severely injured soldiers during the Boer war, in the final quarter of the nineteenth century, and World War One, in the first quarter of the twentieth century (Barnes & Mercer, 2003). It was after World War I in particular that governments and charitable institutions increasingly invested funds in the rehabilitation and retraining of adults blinded in conflicts, and provided resources, protection and jobs to those who were regarded by a grateful nation as returning heroes. This resulted in homes for former soldiers who were blind, by established local and national institutes for the blind. In them they were not just put to work, as child students were in previous centuries, but rehabilitated, and became retrained to return to what was symbolized as a normal life, rather than a deliverance from immorality (le Cue, 1992; RBAS, 1993).

The British disability theorist R. F. Drake (2001) found that such social changes in the early period of the twentieth century also brought about a new focus in Western society on medical conditions, and to the widespread introduction of health and social services. In addition, he also suggests that attitudes about this and many other social issues relating to disability changed during the later decades of the twentieth century, as the global media coverage of the effects of further international conflicts became more accessible. In the twentieth century there was also a medical development that had an enormous impact on how blind people were regarded morally by their sighted peers. The discovery of penicillin in the early years of this century, and its widespread use from the middle of the century provided an effective treatment for conditions which had previously been the main causes of blindness (Del Tufo, 2005; Stanier et al., 1987).

As a result, sexually transmitted disease became less synonymous with blindness, and war injuries became more widely regarded as its cause. This had a major impact on broader social and cultural attitudes to people who were blind in many Western societies, leading to a change in the image of the greater population of the blind people from being street begging indolents of lesser social status to committed subjugators who should be regarded as equals. Injuries suffered by soldiers, sailors and airmen were also mainly the result of multiple injuries, which led to individuals often having several forms of disability which had been previously only rarely linked. Therefore, disability and handicap as a social concept became more important than blindness as an individual concept, and blind people became increasingly known as members of a *handicapped* or *disabled community* more than a *blind* one.

The British disability theorists T. Shakespeare and N. Watson (2001) and C. Barnes and G. Mercer (2003) have also suggested that this was not the only change to the image of people who were blind or had other disabilities. They have suggested that as a result of the growth of mass media, by the end of the Vietnam War attitudes changed even more radically toward those who were regarded as marginalized by Western societies, not just disabled people. This had a profound effect on attitudes toward people who had restricted employment opportunities and/or were economically and socially excluded, as civil rights took on a global dimension. In this way, the disabled population became less known as those who should be pitied and more as those that had been unfairly oppressed. As Barnes and Mercer described:

> From the 1960s, with the gathering forms of social protest spurred initially by the black civil rights movement in America, residential institutions provided a fertile seedbed for disabled activism. The emergence of the independent living movement in America gained a further significant stimulus with the return of disabled veterans from the Vietnam War. (Barnes & Mercer, 2003, p. 23)

In the decade following the Vietnam War, the change in attitudes toward the education of people with disabilities promoted a call for legislation in the United States. These also appeared to arise from the independent living movement, which initiated an academic movement promoting an inclusive pedagogy in the United States. This had then spread to a number of other Western countries by the mid 1970s (Shakespeare & Watson, 2001), such as Canada and Sweden. The effect that this movement had on the formation of an English inclusive movement in education is now examined below.

The Movement Toward Greater Social Inclusion in England

To trace the roots of the 1981 Act in England in particular, it is necessary to start by placing it in the context of previous legislation one hundred years before. In this period, a previous committee of enquiry in the 1860s had informed a government bill, the so-called Forster Bill, in 1870 which gave children with disabilities a legal right to separate education in their local areas. However, the Forster Bill only affected the funding structures of special education (Warnock et al., 1978). Therefore, although important, this and similar legislation in the 1920s, which reiterated this right, had not *informed* the right to an *integrated pedagogy* or the right to an *inclusive education.*

The British educationalist Jonathan Solity (1992) has argued that the modern concept of special needs education, which focused more on psychological disabilities relating to language and behavior, resulted from a notion of measurable intelligences that was introduced into England from France and the United States at the beginning of the twentieth century. As a consequence, he maintained, students who were regarded as incapable of *normal* mental academic performance were segregated, either in separate schools or in streamed classes in their mainstream schools.

Solity continued that this period led to a greater discussion about the nature of segregation and the use of special schools, including those for people who had more recognizable physical disabilities such as blindness. Equally issues of exclusion and segregation based on a notion of social inequality became the focus of investigation during World War II in particular. Accordingly, a review of the whole nature of the then system of education was conducted by the government, first in a green paper published in 1941 and then in a white paper published in 1943. This review was needed, it was felt at the time, because it was foreseen that after the war a significant period of social reconstruction was crucial. It was as a consequence of this review that an important piece of early *inclusive* legislation, which enshrined the right to education for all until the age of fourteen, the 1944 Education Act, was passed through Parliament.

Solity and his fellow British educationalist Catherine Clark (Clark et al., 1997) also suggested that the 1944 Education Act not only had a major effect on the notion of social inclusion in England in the era of social change after World War II it also set out a framework for the contemporary education system that existed until the 1988 Education Act and the introduction of the National Curriculum. In particular, they find that the 1944 Act began a movement that led to a notion of education for all. It also included provision for access to a more mainstream curriculum for children with physical and *mental* disabilities as a national issue. In addition it provided a legal framework for educational provision for *all* children up to the age of fourteen—prior to this act the school leaving age had often differed for children with physical disabilities and for many poorer children. Therefore, at the time it appeared that this legislation represented a major step forward in social as well as educational inclusion.

However, despite its radical reforms Clark and his colleagues found that the 1944 Education Act also formerly encoded the national model of segregated grammar and secondary modern schools, and the notion of setting—where different abilities of children are placed in different classes—within these schools. Furthermore, following the war British educationalists Caroline Benn and Clyde Chitty (1997) alleged that the newly appointed Labour government, who had promised it would work toward reversing this perceived social inequality when it came to power, had not accomplished this and, even worse, had further reinforced segregation among poorer communities. However, Benn and Chitty also claimed that in this period, despite its emphasis on segregated schools, the debate surrounding social disparities drove forward a need for greater social inclusion. In the 1960s, during further social and economic changes, the rise of the comprehensive system also led to questions being raised about other forms of social exclusion, including that caused by *handicap* and *disability*. It was also suggested that this movement provided an opportunity for a further review of the education system, this time to include students with *handicaps* and *disabilities* among its core aims. The formation and implementation of a committee of enquiry from this movement are now examined below.

The Formation of Warnock's Committee

In 1973 Margaret Thatcher, the Secretary of State for Education in Edward Heath's Conservative government, proposed a committee of enquiry to examine the issue of the exclusion of students with *handicaps,* and the need for new legislation to further their education (1978). The then Department of Education and Science (DES) appointed a professor of moral philosophy who specialized in medical ethics, Mary Warnock, as the committee's chair. The DES also recruited members of the committee from existing special schools, academics, members of local boards of education and a lay member whose own child had a disability. As the government had stipulated it, the committee was also assessed by members of the DES and Her Majesty's Inspectorate—the national schools' regulator, who were the guardians of effective teaching and learning in state schools. However, although appearing to be reforming in its inception, the original brief for this committee still contained legacies from the asylum movement. Its central stated purpose was to be training for employment. In addition, the Report also highlighted the efficient financing of separate schools, referred to in its original form as the resourcing of education. Below is an excerpt from the brief illustrating this notion:

> [The committee's purpose is to] review educational provision in England, Scotland and Wales for children and young people handicapped by disabilities of body or mind, taking account of the medical aspects of their needs, together with arrangements to prepare them for entry into employment; to consider the most effective use of resources for these purposes; and to make recommendations. (Warnock et al., 1978, p. 1)

In the early 1970s all disabilities were officially termed *handicaps,* a term which Barnes and Mercer (2003) felt seemed to reflect the charitable nature of the education of people with disabilities. At this time this committee had also been charged with the task of reviewing such terminology and the treatment of these students that resulted from its imposition.[1] As a result a remit was given to the committee of enquiry allowing it to prepare the groundwork for what seemed to be potentially significant changes in the existing educational provision of mainstream as well as special schools. Although Thatcher only stayed in office until 1974 before she became leader of the opposition, Secretaries of State in Harold Wilson and James Callaghan's Labour governments[2,3] and school inspectors[4] were in favor of such reforms. Thus they continued working with the committee of enquiry to provide a radical reform of the education system in much the same way as Thatcher appeared to envisage it. The brief of the original committee was to examine an administrative as well as a pedagogic change within the education system. This would not only affect the curriculum and other pedagogical issues such as teaching provision, but also questioned the structure and philosophy of the separate school system. What now follows is an examination of the functioning of this committee after its formation.

An Examination of the Committee of Enquiry into the Education of Children with Disabilities

Following four years of research, consultation and drafting, the subsequent report was eventually published in 1978 under the Labour Government of Callaghan, when Williams was Secretary of State for Education. As Warnock (Warnock interview, 2000) verified to me, the subsequent *Warnock Report* redefined the notion of groups of children with handicaps as individuals with Special Educational Needs (SEN). The Report chiefly recommended that children with disabilities were to be catered for by an individual assessment of students' subjective needs. In addition, Warnock stated in her interview that there were three further elements of the final report that she felt were of particular importance during its research phase; these were:

(a) [We] wanted to bring home the message that, in the context specifically of *education* (sic.), it did not matter so much about the nature of a child's disability, but on what steps could be taken to ensure that he (sic.) had access, as far as possible, to the curriculum. . . .
(b) There are numerous children with multiple handicaps, and some have disabilities not even classified. . . .
(c) Most important . . . we wanted to widen the scope of 'special education,' beyond those with specific 'handicaps' to include those who, temporarily or permanently, were falling behind their contemporaries, *for whatever reason.* (sic) (Warnock interview, 2000)

At the end of this process and the publication of the report in 1978, Williams (Williams interview, 2000), the then Secretary of State, found that she agreed strongly with the committee's recommendations. More specifically, although she felt that students with severe disabilities could not be catered for in mainstream schools during this period, she believed, on the strength of the report's findings, that those who had less severe disabilities were more suited to mainstream schools. Therefore, the official rhetoric at this time was that the resulting policies were to be for the benefit of all people with disabilities, as this redistribution would allow the remaining resources left in the few remaining special schools. These were becoming increasingly expensive at a time of this economically unstable decade in Britain.

In light of the political movement that was to provide the impetus for educational inclusion, disability rights groups were consulted as a matter of course over the report's findings (Williams interview, 2000). Their feeling was that the purpose of the report was to change mainstream teachers' attitudes toward students with disabilities, persuading them to accept them in their classes. Thus this move toward integration and inclusion was seen as representing a shift in Britain's culturally influenced attitudes toward blindness. Furthermore, this cultural shift was to have an effect on the education of all blind children in England on a similar scale that the introduction of Worcester College had on the small group of middle class students who were blind in the middle of the nineteenth century.

In addition, in order to conform to international best practice a significant proportion of the committee's research was conducted in the United States and continental Europe, as they had not long introduced inclusive policies and laws. As Fish (Fish interview, 2001) remarked, the nature of international society was changing; the rapidity with which it favored the *normalization*, the integration, of people with disabilities, similar to the model being followed in the post-Vietnam era in the US, was escalating. There were, however, further elements that were to interplay with this social movement in England, which seems to have had an effect on subsequent legislation. What now follows in this chapter is an examination of external factors that affected the *Warnock Committee's* report and the 1981 Act.

AN EXAMINATION OF THE PASSING OF LEGISLATION CONTAINED IN THE 1981 ACT

The British educationalists John Visser and Graham Upton (1993) suggested that financial and party political factors affected the design of the British government's attitudes to disability in education, just as they had with earlier institutions for the disabled. Furthermore, the brief and the nature of the research and the findings of the committee were consciously changed to manipulate policies that may have been detrimental to the economic stability of the then government. There is evidence to suggest in the interviews featured in this chapter that the incentive for these changes was in part due to the financial mis-management of previous governments' spending and the need for the incoming Labour government in 1974 to gain power by whatever means. For instance, Fish (Fish interview, 2001) indicated that the terms of reference for the Warnock Committee were not only changed in certain respects after this change of power from a Conservative to a Labour government in 1974, and again from a Labour to a Conservative government in 1979. He also believed that this new Conservative government appeared to reinterpret its main findings before drafting the 1981 Act.

Warnock (Warnock interview, 2000) herself indicated that these changes addressed national issues involving the *financing* of any recommendations. She also said that in the original brief she was similarly disturbed by the restrictions that were placed on her frames of reference. Her original briefing in particular disqualified investigation of what she regarded as vital social and economic issues, such as the linkage between learning difficulties and poverty. Prentice (Prentice interview, 2001) advanced further evidence as to why finances were such a contentious issue. He found that in the early 1970s there was a hidden financial crisis in the British government. In particular he stated that the resources the Conservative government had to spend prior to the election in 1974 appeared to be inaccurately presented to the opposition and the public. Although he felt he could not say whether members of his government were aware of this factor or not before being elected, Prentice did say that the Labour Government came to power in 1974 on the promise of new spending based on an optimistic state of affairs in the British Treasury. This, he continued, was a mistake. Prentice continued with this recollection:

> I can be more eloquent!
>
> Economic pressures had always affected educational funding and always will. Even when rises in spending are available, (i.e. in most years, to some extent) there is generally a larger rise in needs and/or expectations.
>
> In the mid-'70s and early '80s [when the 1981 Special Educational Needs Act was debated and introduced] the pressure was intense. Economic pressures pointed to lower spending.
>
> *The Mid-'70s*
>
> We took office confident that we could reverse "Tory cuts." Following my appointment on the Monday after Polling Day [to Secretary of State for Education], I went to our first cabinet that afternoon. On the way in, Denis Healey, the new Chancellor, told me we must meet soon to [review] some spending plans.
>
> The reality: no extra money and, a few weeks later, deeper cuts.
>
> For example, Margaret Thatcher (my predecessor) abolished free school milk for older children so as to direct more funds to nursery education and to the needs of handicapped children. The Labour manifesto had promised (a) to restore the milk, (b) to expand nursery education much more. It was soon

> clear to me that we could not do both. I rapidly decided to drop the proposal for school milk. A few weeks later I was faced with cabinet decisions for more cuts. Reluctantly I cut nursery provision below the Thatcher level, so as to [fund] the basic primary and secondary programmes.
>
> Strange to say, I achieved one major breakthrough which cost a lot of money—the appointment of the Houghton Committee, and the agreement of the Cabinet, in advance, to implement its proposals. This led to the biggest salary increase in the history of the profession.
>
> (In these days all Departmental estimates were automatically increased for rising costs, including salaries. This is no longer the case.)
>
> Later in the 70s there were more cuts due to the economic pressures, culminating in the IMF crisis in 1976 [in which Denis Healey had to borrow money from the International Monetary Fund to stabilise the British Economy]. (Prentice interview, 2001)

Thus it seems that economic circumstances in part forced the British government to change the brief of the committee to favor making changes that would save the government money, such as reducing the number of special schools, but at the same time avoid issues that would cost it money, such as investing more money into areas where poverty was high. This manipulation allowed them to avoid further spending promises they could not afford in an attempt to secure or gain political power. It also seems possible that the original terms of reference for the committee drafted by the Conservatives could also have affected this cost saving exercise. In particular, moving toward inclusion without the expensive consideration of social deprivation and disability, the provision for which was left out of the enquiry by Thatcher's ministry, meant that inclusion required less financial support than maintaining special schools.

This need for financial and political expediency did not finish with the publication of the committee's report in 1978 or the change of government in 1979. Evidence from the interviews (Williams interview, 2000; Prentice interview, 2001) suggests that although it may have been prudent to bring forward the drafting of a special needs act, its drafting was delayed until the Conservative government took over in 1979. As Williams argued, the Labour government in the mid to late 1970s had too small a majority to carry through a bill of their own in the face of this opposition from certain elements of the Conservative party. This was particularly in respect to the inclusion of students with disabilities and learning difficulties in mainstream schools, which it was felt at the time affected the education of more *able* children. Thus, this problem not only affected the drafting of a bill under the then Labour government, it also had its effects on the subsequent 1981 Act.

This segment now turns to an examination of these factors.

The Political Process Leading to the 1981 Act

In 1979 the Labour party lost the general election, and the Conservative party gained office under the leadership of the former Secretary of State for Education, Margaret Thatcher. In the same year, Reginald Prentice crossed the floor of the House of Commons, resigning his post in the Labour party and taking a new role in the new Conservative government. He was duly allowed to maintain his position as an MP in the new government, and also accepted the post as Minister of Disability in the subsequent Thatcher cabinet. Over the next two years, Prentice (Prentice interview, 2001) and Fish (Fish interview, 2001) alleged, the 1981 Act was rushed through Parliament and into law without many vital details, such as staff training and considerations of the ability of mainstream schools to cope with students with disabilities, being scrutinized. Three factors were raised which particularly influenced this haste:

- The first factor was financial—Fish and Prentice found that funding was purposely not cited as an element of the law.
- The second was international politics—Prentice found that the Act had to coincide with the United Nations' International Year of the Disabled in 1981.
- The third was the belief that inclusion of people with disabilities would prevent a rise in educational standards—both Fish and Prentice felt that the findings of the report clashed with the aims of the Conservatives' promise to raise standards in schools through later education acts.

Furthermore, Warnock (Warnock interview, 2000) felt that the negative effects of this lack of consideration and the speed of the introduction of the 1981 Act led to many of her recommendations being ignored (Warnock interview, 2000; Visser & Upton, 1993). In particular, in her interview she stated that the categorization of special needs led to a further re-classification of disability, based on the severity of individual disabilities—that eventually became more detrimental to those students classified under this system than before the 1981 Act. For instance, in the new classification system people were not categorized as having a visual impairment; they were classified between a stage one educational need or a full statement—this was termed a *stage five special need*, on a sliding scale where one was weakest and five was highest. This could have been avoided in the drafting of the Act, Warnock argued, with more time and consideration being provided.

This chapter now finishes with a conclusion of our findings.

CONCLUSION

This chapter has shown that the political process of implementing the 1981 Act was at least partly motivated by social and cultural considerations, such as the interlacing and manipulation of promises to win elections. In addition, in a way that was similar to the introduction of private asylums in the Eighteenth and Nineteenth Centuries, it could be suggested that social factors such as cultural pressure and the changes in perceptions of the causes of blindness also affected a change to the motives of the civil rights movements in Britain when it came to educational inclusion. These arguments are now considered in more detail underneath separate subheadings.

Changes in Attitudes to Blindness Caused by War

This chapter demonstrated that the perception of the major cause of blindness had an effect on charitable notions of blindness. In particular, in the twentieth century there was a gradual progression from a perception of *the immoral blind* to the image of *the heroic blind.* As a result, rather than concentrating on institutions that were designed to reform immorality after the return of many people who were blinded in major wars, the focus of institutionalization of people who were blind after this period focused on a *return to normality* of those who were the social peers of sighted people. This in turn greatly altered cultural notions of what it was like to be blind in England. This is not to say that the whole of society changed their cultural attitudes toward *the blind* as a community, but that there was a definite change in the institutional attitudes about the late blind, and a changing shift in the beliefs about the psychological condition of blindness in particular.

Furthermore, as the end of the twentieth century approached an increase in access to the media, particularly the electronic media, helped others to understand the notion of disability formed by war. The first prominent war to be featured on color television in particular, that of Vietnam, raised public awareness of the issues of fit young men, with many years of work and *normal* life ahead of them, returning with disabilities. This in turn led to sympathy and support for the independent living movement in the United States, influenced new laws in North America and continental Europe, and became a focus of the Warnock Committee's later investigation. However, in England this background to an understanding of disability rights was only a part of the overall picture. Evidence presented in this chapter also shows that financial and political incentives had an impact on the notions of legal inclusion. This issue is now discussed under the subheading below.

Financial Considerations in the Implementation of the 1981 Act

There is a significant linkage between the research and passing of legislation that formed the 1981 Act and economic circumstances in Britain in the 1970s and early 1980s. First, circumstantial evidence suggests that the original notion of developing a policy of inclusion was motivated in part by financial hardship in England, rather than the notion of the augmentation of civil liberties. For instance, in her original brief, Warnock was specifically told that she could not include socio-economic considerations. As a result she was restricted to findings based purely on practical needs for facilitating inclusion, rather than addressing issues as to whether special needs, such as disability, were caused by deprivation. This stipulation, originally introduced by the Conservatives but left unchanged by the subsequent Labour government, recommended that resources be diverted to overcome these causes.

Furthermore, when Prentice entered government in the mid-1970s as a member of the cabinet, his budgets were severely restricted. This also appeared to be the case when he was made a minister in the late 1970s and early 1980s Conservative administration. In particular, in the mid 1970s he was informed by the then Chancellor of the exchequer, Denis Healey, that the spending promises the Labour government had made needed to be curtailed. The only promise upheld in this era was a teachers' pay rise, which was an issue they had discussed with unions beforehand. As a result, the 1981 Act was passed without due thought to the impact it would have on existing mainstream schools (who had no extra resources from central government in this period to help with inclusion), no extra training provision for mainstream teachers—even though this was recommended by the committee and disability groups—and no extra consideration for the social support of students with disabilities from poorer backgrounds. We continue by describing the affects of political issues on the terms and implementation of the 1981 Act.

Political Motivations Affecting the 1981 Act

There is a significant linkage between the haste in passing the 1981 Act and further party political factors. First, in the 1970s, as Prentice described in his interview, the Labour party promised to increase its resourcing of support for students with disabilities in education, either unaware of the true finances available or aware but not giving a truthful picture of what could be done with them. The only promise that was kept was that of a teachers' pay rise, negotiated with a union which politically and financially supported the party and who had helped with their election campaign. In the lead up to the 1981 Act, their frames of reference also changed with each change of government. Therefore, the research conducted by the committee in this period would have been severely restricted by such changes, and important issues either ignored or reduced in the final report; the impact of social deprivation on disability, stated by Warnock, was just such an issue.

Finally, as Prentice mentioned, the political incentive of coinciding legislation that promoted the rights of people with disabilities and the 1981 International Year of the Disabled led to the legislation being hurried through parliament. Both Fish and Warnock's opinions appear to concur with this finding, as they believe that many recommendations made by the committee were ignored or unwisely considered in this haste.

What now follows is a conclusion to this book's findings, and further recommendations.

NOTES

1. *Warnock interview*: Postal interview with Baroness Mary Warnock, Chair of the Committee of Enquiry, HM Government, 1974–1978. Interview answers received in May 2000.
2. *Prentice interview*: Postal interview with Lord Reginald Prentice, Former Secretary of State in the Labour Government of Harold Wilson. Interview answers supplied in February 2001.
3. *Williams interview*: Postal interview with Baroness Shirley Williams, former Secretary of State for Education in the Labour Government of James Callaghan. Answers received in January 2000.
4. *Fish interview*: Postal interview with John Fish, Deputy Chief Inspector of Schools involved with the Warnock Report and the 1981 Special Education Needs Act. Interview answers supplied in June 2001.

PART III

CHAPTER 7

CONCLUSION

> What is the true object of education? Should the educator aim at training the largest possible number of individuals to be of the greatest service to the state, up to the limit of the capability of each, or should he rather try to give each one an opportunity to develop fully the best qualities which he possesses, regardless of whether this method of training may or may not seem to the immediate practical use either to the person or to the community? The question, in most ages and most countries, does not admit of a simple answer. (Dobson, 1963, p. 1)

INTRODUCTION

In this final chapter, we conclude our discussion on blindness and touch. Its aims are to summarize our findings and identify future directions for research and debate. This chapter is divided into two sections. The first looks at the processes of change in English attitudes to blindness and touch discussed in the introduction and title of this book. The second and final section develops social, cultural and political arguments that arise from the debate of our discussion, and makes recommendations for the future.

This chapter now continues by discussing the initial hypotheses of this thesis in light of the evidence presented in the study.

God, Money, and Politics, pages 109–116

SOCIAL AND CULTURAL INFLUENCES ON THE DEVELOPMENT OF ATTITUDES TO BLINDNESS IN EDUCATION

As we suggested in its title, this book has shown, how three major external social and cultural factors affected our understanding of blindness and touch in England: the first is the theological debate evident in the literature that influenced English attitudes and educational institutions; the second is the need for financial sustenance of individuals and institutions; and the third is political manipulation by educators, academics and government ministers in order to gain or maintain their political power—or covertly influence a greater national debate on the power of monarchs and religious institutions. This has led to a clear chronology of social and cultural development, which shaped attitudes to blindness in England. These can be split into three distinct phases of history: attitudes affecting the asylums, attitudes affecting schools and colleges, and attitudes affecting integration in education.

These three phases are now discussed under separate subheadings.

Attitudes Affecting the Asylums

The initial debate that would later inspire the foundation of institutes for the blind, whether as asylums or schools, was founded on religious politics. It was, in particular, motivated by a need to question the authenticity of received theological wisdom by Locke and his circle of philosophers in the latter years of the seventeenth century. Although our discussion only provided circumstantial evidence that this was directly related to more in-depth Gnosticism, such as Newton's debate with Locke, there is enough evidence from Locke's writing itself to confirm that his theory of blindness and touch was directly related to theological notions of innate ideas, and its presupposed relation to vision. This led to a strand of academic thinking based on the study of blindness and touch that was often quoted by later academics and philanthropists, who were to form the first institutes for the blind.

Religious, political and financial factors were also instrumental in informing the initial observation that blind people were often found begging in wealthy cities, that sexually transmitted diseases caused a sizeable proportion of the blindness in Europe before the twentieth century, and thus the resulting view throughout Western culture that working and underclass blind people had different moral and ethical standards to those with sight. Again with the support of notions put forward in Locke's essay that gathered momentum and fashionable supporters in the eighteenth century, this led to a cultural backlash during the later periods of the Enlightenment. In this era, blindness, and by extension *the blind person*, were held up as a symbol of conformity, anti-establishmentarianism, moral purity or as an object of disgust, laziness, economic inefficiency and consequently immorality—this movement was so successful in neighboring France that, as authors such as Jay and Paulson commented, sight was actually derided as a perception of the establishment.

For example, the radical author Diderot felt that blind women were, in many cases, morally superior and sexually more innocent because of their blindness; while educators such as Demodocus, Hauy, Rushton, and Bath and Fox, whose notions inspired the first curricula in England, felt blind people should be pitied for their indolence and moral inferiority, caused by a lack of perception of beauty, and a lack of exercise caused by the inability to work or learn. Consequently, the first asylums were built in England to *cure* what were felt to be the worst social and irreligious effects of blindness, such as begging and laziness, by introducing mechanical and industrial handcrafts; this perceived immorality offended the religious and social morals of those founding the asylums so much that Rushton, for instance, stated that this indolence was the worst curse that could be inflicted on those who were young and blind.

Unlike their continental counterparts, English asylums were also unique in the nineteenth century as they were self funding, and received no government assistance. As a result their emphasis on a commercial handcrafts curriculum reflected this moral foundation, which promoted a need to lift people who were blind out of poverty and prevent laziness and indolence. In addition, sales of the products produced by this form of labor and derived from long working hours also helped to fund the asylum's maintenance and administration.

We now reflect on influences on schools and colleges for the blind.

Attitudes Affecting Schools and Colleges

The religious beliefs of the founders of English colleges and later schools for the blind also influenced a notion that people who were blind would benefit most from an aesthetic *oral* culture rather than an aesthetic *tactile* culture. Although in the early twentieth century schools for the blind began to believe that people who were blind could appreciate artifacts aesthetically through tactile perception, before this period the curriculum of many of these schools reflected a belief that tactile perception was only able to transmit information, such as Braille literature or three dimensional representations of clay flowers, not aesthetic appreciation.

Furthermore, this was reflected in a broader academic culture, as the existence of a tactile appreciation of an aesthetic art form was not widely understood by science until the latter quarter of the twentieth century; this was despite its practice in isolated instances in Austria from the beginning of the nineteenth century. For example, before this period several psychological studies of blindness, such as those of von Senden, doubted whether people who were blind could appreciate the fine arts through touch. Similarly it also appears that educational projects, such as those of Klein and later Lowenfeld, remained un-replicated in schools in England, even though its representative organization, the College of Teachers for the Blind, was beginning to recognize the use of handcrafts as a mode of emotional development from the 1950s.

Money and politics also had an influence on the treatment of different social classes of blind students. Again the tactile skills of people who were blind were first discussed by writers such as Locke and Diderot, who observed the skill of English blind men and the musical dexterity of French blind beggars in order to demonstrate their mental *normality*. However, in common with the originators of the asylums, the founders of the original schools and colleges also felt that this observation could be used as a form of moral and vocational education for poor underclass and working class people who were blind in their area, to lift them out of what was perceived as a moral poverty. The observations in the later literature by the founders and theorists of the first asylums in England identified manual labor and musical crafts in particular as arts that could be used by people who were blind to worship God, and to derive trades such as cane weaver, church organist or piano tuner. Consequently it also became the most important element of later English curricula of charitable schools and colleges, such as the Normal College. Again, however, this form of music training was culturally distinct from that practiced earlier in France, which relied on Braille notation in writing and in music, and therefore appeared to rely on a literary and oral aesthetic culture as its basis of education.

Furthermore, the founding of Worcester College, charged with the education of middle class students, illustrated a political and economic gap in the teaching of music and mechanical arts as an aesthetic art rather than simply a mode of labor. In particular, the curriculum presented at Worcester College appeared to spurn all forms of handcraft, and focus on instrumental teaching. This assumed a high degree of musical appreciation, performance and composition. Hence, the students at Worcester College—before it was administered by the RNIB—received a higher standard of academic musical education than their counterparts in France had at the end of the eighteenth century. In addition, through their curriculum, they were prepared for entry to prestigious universities and colleges of music, rather than merely prepared for jobs with low pay.

We now discuss the influences to attitudes toward educational integration.

Attitudes Affecting Educational Integration

Financial and political expedience were very much in evidence during the latter quarter of the twentieth century in the research and drafting of the 1981 Act, the law that provided integration for blind students in mainstream schools. In the process of drafting this legislation evidence indicates that financial matters were manipulated in order to save money for the British economy, which was in severe difficulties at the time. This led to the elimination of investigation into the social and economic disadvantages of students who were blind and excluded from the brief of the Warnock Commission, and to the exclusion of expensive training and resource provision of mainstream schools from the drafting and implementation of the 1981 Act.

In terms of the political manipulation of this law, evidence presented by Warnock and Prentice indicates that spending promises such as those given for the education of students with disabilities, were purposely manipulated by the Conservative and Labour parties in the 1970s to either gain or maintain power. Furthermore, Prentice explained that the drafting of the 1981 Act was too hurried as it had to be presented as part of Britain's contribution to the International Year of the Disabled. As a result, it would seem that many important elements, such as student teaching support, training and social integration, were again ignored.

What now follows are further conclusions and recommendations based on the general theme of these findings, and personal observations made during the research process.

PERSONAL OBSERVATIONS AND RECOMMENDATIONS

> Martin [Luther] King was right: "The hour is late" and "the clock of destiny is ticking out" . . . Human beings are meant for life and not death. They are meant for freedom and not slavery. They were created for each other and not against each other. We must, therefore, breakdown the barriers that separate people from one another . . . [For] all who have given their lives in the struggle for justice, let us direct our fight towards one goal—the beloved community of mankind. (Cone, 1991, p. 318)

My original study of blindness and touch began more than fifteen years ago, as a pedagogical investigation of adults in an extra mural art class in Leicester. The room where these classes were held was an old storage cellar for the student bar at Leicester University. To get into it, you had to weave a chicane through beer barrels or climb one steep set of stairs, and descend another once in the room. The floor was very rough and the equipment and students' work had to be kept in little corners as no proper storage was provided. It was totally inappropriate. But it immediately taught me something fundamental about the way that people who were blind or had other disabilities were regarded by these institutions.

The more I conducted the study, the more I realized that the single biggest factor influencing education—classroom practice, assessment, school and college policy, and so forth—were the moral, political and financial motivations that occurred above the heads of both teachers and students. This even influenced classroom practice. *Pedagogy was a political issue*, I found; I believed it then and when I worked with[1] children with special educational needs I began to realize it more. This, I would argue, is a significant problem with disability in education. It is run from the ideological viewpoint of people who dictate policy or assert what people with disabilities should learn. These institutes control definitions of disability and policies from the center. They are moreover fundamentally *undemocratic*.[2] It is too often the model of pedagogy that is, unfortunately, subsumed in this malaise.

My broadest conclusion to the epistemology of blindness, therefore, is that we need a new way of thinking about *it*, not just a new way of *teaching* people who are blind and visually impaired. We should not just rearrange the thoughts we have at the moment and as we have throughout history. Recommending pedagogic strategies or policies would be useless without such reforms. On this point Mary Warnock and John Fish argued from an astute and experienced position: we need a new way of classifying individual students, their abilities and disabilities, and most important their needs. I would also extend this argument to all *socially* classified groups. Our old, outmoded models of study, which simply shift definitions and cause new problems, are most to blame for exclusion. In other words, it is not the mode of classification that is the problem; it is the need to classify in the first place

I would argue this is not a naïve idea, although perhaps it has its own dangers of becoming an ideological one. It should also allow us as a society to attempt to avoid the mistakes made with the classification of race. In particular, we should learn from the observations of Martin Luther King before his assassination, and discussed in the quote above by the American writer on the history of racial issues, James H. Cone. King's understanding came out of his experiences away from a formally segregated society at universities and in his later vocations. It also came from his observation that in the United States people in extreme poverty were White as well as Black—a common fallacy in both *racial* communities.

As Cone argues, King's political activities and rhetoric were slowly being led away from a notion of racial politics to one in which race was a social trait among many. This led him to campaign and preach on more general social issues, such as workers' rights[3] and the necessity of proper health and education for *all people*, of *all colors*. In this, Cone continued, he appeared to be more influenced by the politics of Ghandi, in which the good of the most in need could be seen in the context of provision for all. The quote below illustrates this point:

> Martin's interest . . . was motivated [amongst other factors] . . . by the failure of the civil rights acts and President Johnson's War on Poverty to affect significantly the life chances of the black poor in the urban ghettos of America. The Watts riots and the many others that followed it revolutionized Martin's thinking on America. He began to see that the problem of injustice in America was deeper and wider than racism. . . .
>
> Martin's Poor People's March to Washington was intended to achieve solidarity among the poor of all races. Perhaps it would indeed have been a powerful symbol of the solidarity of the poor, if he had not been assassinated while supporting garbage workers in Memphis. . . . If we today are to complete what Martin . . . failed to achieve, then we must identify and correct [this] weakness. (Cone, 1991, pp. 286–287).

If we could apply such a notion to disability, it would become less an emotive issue but a practical one. By applying this approach to such traits, we can regard what are subjective (Hayhoe, 2005b) causes of disability in a manner that eliminates practical difficulties experienced by individuals, not classifications of people. This was arguably what the 1981 Act was and is trying to achieve. Such an approach would also allow us to contextualize measures to alleviate disabilities, rather than regarding them as broad political issues or whole person problems. In particular, in this topic these measures could include seemingly small changes, such as a greater awareness of the uses of German film, the nature of Braille, the need for large type, or the classroom management strategies designed for students needing to do close-work.

Furthermore, I would argue that we need new commissions to alleviate the mis-classification of special needs. Again, I am in agreement with Baroness Warnock in believing that it is time for such new committees of enquiry to address the needs of all children, and they should not just ring fence, label or stigmatize a small minority of people. Perhaps most important, however, in doing this the government has to see education as a pedagogic not a political issue. It would be unthinkable for government departments to prescribe how a general practitioner should diagnose chicken pox or flu, and so why should it treat the expertise of practicing, *trained* teachers in the same way; when they are properly informed and given access to appropriate experience? They more often have the best diagnosis of what is suitable for children under their care.

Thus, all of these recommendations come to naught if the most fundamental change is not achieved. All teachers, school workers, parents and students themselves need to be educated properly in the differences and nature of physical traits that can lead to disabilities. It should not just be a chosen few special needs teachers and theorists who are provided with this understanding; it is everyone who influences the day to day lives of all children. Disability is not just a whole school problem; it is a whole society problem. Not until there is the cultural and social will in governance and the community through such training and awareness to alleviate inequity in the contexts where it exists, will there be a sea change in real teaching and learning. For this, I hope, like Cone and King, we don't have to wait too long.

NOTES

1. In this period, I began key-working students with special needs as well as teaching them. This title in English classrooms is given to someone who is the main point of contact for the child in the school in terms of their special needs. They are also the point of contact for staff that either have problems or need information or advice about the child.
2. In this context, I use the term in the original Greek meaning "of the people."
3. Martin Luther King, Jr., was murdered while campaigning for the rights of sanitation workers in Memphis.

BIBLIOGRAPHY

Barasch, M. (2001). *Blindness: The history of a mental image in western thought.* London: Routledge.

Barnes, C., & Mercer, G .(2003). *Disability.* Cambridge: Polity Press.

Bartlett, J. E. (1955, February). Museums and the blind. *The Museums Journal, 54,* 283–287.

Benn, C., & Chitty, C. (1997). *Thirty years on: Is comprehensive education alive and well or struggling to survive?* London: Penguin.

Benson, K. R., & McNeice, W.C. (1964). *Crafts for [the] retarded.* New York: McNight & McNight.

Berkeley, G. (1899). *Selections from Berkley.* (A. C. Fraser, Ed., 5th ed. amended). Oxford: Clarendon Press.

Bindman, D. (1981). *Hogarth.* London: Thames & Hudson.

Bristol Asylum. (1800). *The state of the Bristol Asylum For the Blind from the 31st December 1798 to the 31st December 1799.* Bristol: Bristol Asylum.

Bristol Asylum. (1838). *Appeal to the Public by the Bristol Asylum, or the School of Industry for the Blind.* Bristol: Bristol Asylum.

Bristol Asylum, the Honorary Secretary. (1854). *An historical sketch of the Bristol Asylum For the Blind.* Bristol: Chapman & Hall.

Bristol School for the Blind. (1887). *Annual report of the Bristol School for the Blind.* Bristol: The Bristol School for the Blind.

Bristol School of Industry for the Blind. (1908). *Exhibition and sale of work.* Bristol: Bristol School of Industry for the Blind.

Campbell, F. J. (1901). *National education and training of the blind.* London: The Royal Normal College & College of Music for the Blind.

Charlton Deas J. A. (1913, September). The showing of museums and art galleries to the blind. *Museums Journal, 13*(3), 85–106.

Clark, C., Dyson, A., Millward A. J., & Skidmore, D. (1997). *New directions in special education.* London: Cassell.

Coakes, R. L., & Holmes Sellors P. J. (1992). *An outline of ophthalmology.* London: Butterworth-Heinemann.

Committee of the College of Teachers for the Blind. (1956). *Handbook for school teachers of the blind.* Worcester: College of Teachers for the Blind.

God, Money, and Politics, pages 117–123

Cone, J. H. (1991). *Martin & Malcolm & America: A dream or a nightmare.* Mary Knoll, NY: Orbis Books.

Craze, M. (1972). *King's School Worcester.* Worcester: Ebenezer Baylis & Son Ltd.

De Coster, K., & Loots, G. (2004). Somewhere In between touch and vision: In search of a meaningful art education for blind individuals. *The International Journal of Art & Design Education, 23*(3), 326–334.

Del Tufo, B. (2005). *Aesthetic experience and blindness.* Unpublished doctoral dissertation, Middlesex University, England.

Demodocus (1774, October). On the education of the blind (open letter, dated September 10th 1774). *The Edinburgh Magazine and Review, 2*, 673–686.

Dharma Colwell, S. (1993). *The education of adults with disabilities.* Unpublished master's thesis, Bristol University, England.

Diderot, D. (2001). Letter on the blind for the use of those who see. In *Thoughts on the interpretation of nature and other philosophical works.* (M. Jourdain, Trans.). Manchester: Clinamen Press.

Dobson, J. F. (1963). *Ancient education and its meaning to us.* New York: Cooper Square Publishing Inc.

Dobson, J. (1982). *Touch perception.* Unpublished graduate diploma in education, Nottingham University, England.

Dodds, A. (1980). *Spatial representation and visual impairment.* Unpublished doctoral thesis, Nottingham University, England.

Drake, R. F. (2001). Welfare states and disabled people. In G. L. Albrecht, K. D. Seelman, and M. Bury (Eds.), *Handbook of disability studies* (pp. 412–438). London: Sage.

Eriksson, Y. (2003). What is the history of tactile pictures? In E. Axel & N. Levent (Eds.), *Art beyond sight.* New York: The American Foundation for the Blind & Art Education for the Blind.

Feeney, D. (2004). *Towards an aesthetics of blindness: An interdisciplinary response to Synge, Yeats and Friel.* Unpublished doctoral thesis, Trinity College, Dublin, Ireland.

Feinberg, E. B. (Ed.). (2003). *US national library of medicine* (encyclopaedia). Bethesda, MD: National Library of Medicine. Featured on http://www.nlm.nih.gov.

Fergusson, B. (1987). *The value of museums to visually impaired children and young people with particular reference to several museums in Yorkshire.* Unpublished thesis, Birmingham University, England.

Fletcher, R. C. (1984). *The college on the ridge.* Worcester: The Trinity Press.

Foucault, M. (1989). *Madness and civilisation.* (R. Howard, Trans.). London: Routledge Classics.

Gabias, P. (2003). A researcher who is blind, on tactile pictures. In E. Axel & N . Levent (Eds.), *Art beyond sight.* New York: The American Foundation for the Blind & Art Education for the Blind.

Glanville-Smith M., Cross G., Brookes A. & Hughes H. (1982, June) *South East Worcester Parish Team News.* Worcester: Worcester Diocese.

Goffman, E. (1990). *Stigma: Notes on the management of spoiled identity.* London: Penguin.

Goffman, E. (1991). *Asylums.* London: Penguin.

Gregory, R. L. (1974). *Concepts and mechanisms of perception.* London: Duckworth.

Gregory, R. L. (Ed.). (1987). *The Oxford companion to the mind.* Oxford: Oxford University Press.

Groce , N. E. (1988). *Everyone here spoke sign language: Hereditary deafness on Martha's Vineyard.* Cambridge, MA: Harvard University Press.

Guillie. (1819). *Translation of an essay on the instruction and amusements of the blind.* London: Sampson, Low, Marston.

Harris, R. W. (1979). *Aesthetic development of visually impaired children: A curriculum model for grades kindergarten thru six.* Unpublished doctoral thesis, Indiana University, Bloomington, IN.

Hauy, V. (1889). *An essay on the education of the blind* [Reprinted from the original from Paris, 1786]. Leicester, England: Tompkin & Shardlow, St. Martins.

Hayhoe. S. (1995a). *The art education of blind adults.* Unpublished masters thesis, Leicester University, England.

Hayhoe, S. (1995b, September). *The history of art education of blind students.* European Conference on Educational Research, Bath University.

Hayhoe, S. (1995c). *Arts education and blind students.* Educational Research Students Network Conference, Birmingham University.

Hayhoe S. (2000). The effects of late arts education on adults with early visual disabilities. *Educational Research & Evaluation, 6*(3), 229–239.

Hayhoe, S. (2003). The development of the research of the psychology of visual impairment in the visual arts, In E. Axel & N. Levent (Eds.), *Art beyond sight.* New York: The American Foundation for the Blind & Art Education for the Blind.

Hayhoe, S. (2004, April). *The development of an epistemological model of disability: The Enlightenment, scientific theories on blindness and art/craft education in the US.* Guest Lecture delivered in the Psychology Department, University of Toronto at Scarborough.

Hayhoe, S. (2005a). *The Four Senses Project (including the Four Senses manifesto).* Paper presented in support of The Four Senses Exhibition, The Henry Moore Gallery, Royal College of Art, London during a Guest Lecture for Art Through Touch, London, sponsored by the Art Council.

Hayhoe, S. (2005b). *An examination of social and cultural factors affecting art education in English schools for the blind.* Unpublished doctoral thesis, Birmingham University, England.

Hayhoe, S. (2006, February). *How Locke, Molyneux and the Enlightenment affected the way we think about the arts and blindness, and what is the way forward?* Guest lecture given at the seminar "The Fifth Sense" for the government of the Lombardi Region, Milan, Italy.

Heller, M. (2003). How do blind people perceive and draw pictures In E. Axel & N. Levent (Eds.), *Art beyond sight.* New York: The American Foundation for the Blind & Art Education for the Blind.

Hinton, R. (1991, November). First introduction to tactiles. *The British Journal of Visual Impairment, 9*(3). http://www.visugate.biz/bjvi/1991/nov1991.html.

Hopkins, R. (2000, January). Touching pictures. *British Journal of Aesthetics, 40*(1), 149–167.

Hopkins, R. (2003, March). *Painting, sculpture, sight and touch.* Professorial Lecture delivered at Sheffield University, England.

Howe, S. G. (1833a). Address of the trustees of the New England Institution for the Education of the Blind to the public. Boston: Carter, Hendee.

Howe, S. G. (1833b, July). The education of the blind. *The North American Review, 37*, 20–55.

Howe, S. G. (1836a). Art.VIII. Education of the blind. *The Literary and Theological Review, III/-.*

Howe, S. G. (1836b, August). *The education of the blind.* Guest Lecture delivered to the American Institute of Instruction, Boston.

Howe, S. G. (1837). *A discourse on the social relations of man.* Boston: Marsh, Capen & Lyon.

Hull, J. (2003). The world of sight and the world of touch. In E .Axel & N. Levent (Eds.), *Art beyond sight.* New York: The American Foundation for the Blind & Art Education for the Blind.

Hull, J. M. (2001). *In the beginning there was darkness.* London: SCM Press.

Hunter, R. (2002). *Forgotten hero: The life and times of Edward Rushton, Liverpool's blind poet, revolutionary republican and anti-slavery fighter.* Liverpool: Living History Library.

Illingworth, W. H. (1910). *History of the education of the blind.* London: Sampson Low, Marston.

Jansson, G., & Holmes, E. (2003). Can we read depth in tactile pictures? In E. Axel & N. Levent (Eds.), *Art beyond sight.* New York: The American Foundation for the Blind & Art Education for the Blind.

Jay, M. (1994). *Downcast eyes: The denigration of vision in twentieth-century French thought.* Berkley: University of California Press.

Joint Committee of the College of Teachers of the Blind (& the National Institute For the Blind). (1936). *The education of the blind.* London: Edward Arnold.

Jones, W. D. (1970). *Twenty four selected articles* (F. J. H. Tooze, Ed.). Worcester: College of Teachers for the Blind.

Kennedy, J. M. (1974). *The psychology of picture perception.* New York: Jossey-Bass.

Kennedy, J. M. (1983). What can we learn about pictures from the blind? *American Scientist, 7*(1), 119–126.

Kennedy, J. M. (1993). *Drawing and the blind.* New Haven, CT: Yale University Press.

Kennedy, J. M. (1997). How the blind draw. *Scientific American, 276*(1), 60–65.

Kennedy, J. M., & Merkas, C. (2000). Depictions of motion devised by a blind person. *Psychonomic Bulletin and Review, 7*, 700–706.

Kinsbourne, M., & Lempert, H. (1980). Human figure representation by blind children. *Journal of General Psychology, 102*, 33–37.

Klein, J. W. (1971). Guide for a suitable treatment of blind children from their earliest youth on in the circle of their families and in the schools of their home communities. *Blindness, 1971*, 231–242.

Kuhn, T. S. (1996). *The structure of scientific revolution* (3rd ed.). Chicago: University of Chicago Press.

Lawson, J., & Silver, H. (1973). *Social history of education in England.* London: Methuen Young Books.

le Cue, J. (1992). *A time of transition: BRSB the first two hundred years.* Bristol: Bristol Royal Society For the Blind.

Liverpool School For the Blind. (1849). *Report of the state of the School For the Blind, Liverpool, from the 31st December, 1847, to the 31st December, 1848.* Liverpool: George McOrquodale.

Locke, J. (2001). *An essay concerning human understanding* (J. W. Yolton, Ed.). London: Everyman Library/J.M. Dent.

Lowenfeld, B. (1981). Effects of blindness on the cognitive functioning of children. In B. Lowenfeld, *On blindness and blind people: Selected papers.* New York: American Federation for the Blind.

Lowenfeld, V. (1951). *Psycho-aesthetic implications of the art of the blind.* New York: American Foundation for the Blind.

Lowenfeld, V. (1959). *The nature of creative activity.* London: Routledge & Kegan Paul.

Lowenfeld, V., & Brittain, W. L. (1987). *Creative and mental growth* (8th ed.). New York: Macmillan.

Lowenfeld, V., & Munz, L. (1934). *Sculpture for the blind.* New York: R. M. Rohrer.

Magee, B., & Milligan, M. (1998). *Sight unseen.* London: Phoenix.

Micheal, J. A. (1981). *Viktor Lowenfeld: Pioneer in art education therapy.* New York: American Federation for the Blind.

Millar, S. (1975). Visual experience or translation rules? Drawing the human figure by blind and sighted children. *Perception, 4,* 363–371.

Mülheis, V. (2004). *Kunst im Sehverlust: Eine phänomenologische Studie.* Unpublished doctoral thesis, University of Köln, Netherlands.

Nagel, T. (1991). *Mortal questions.* Cambridge: Canto.

Newton, I. (1995). *Newton correspondence.* (I. B. Cohen & R. S. Westfall, Eds.). New York: WW Norton.

Newton, I. (1959–1977). *The correspondence of Isaac Newton.* (H. W. Turnbull, J. F. Scott, A. Rupert Hall, & L.Tilling, Eds.) (Vols. 1–7). Cambridge: Cambridge University Press. From the Correspondence of Isaac Newton, Portsmouth Collection, Cambridge University Library, UK.

Pablasek, M. (1873). *Report of the First Congress of Teachers of the Blind.* (B. Lowenfeld, Trans.). Standing Congress Committee of Vienna, Austria.

Paulson, W. R. (1987). *Enlightenment, romanticism and the blind in France.* Princeton, NJ: Princeton University Press.

Pearson, A. (1991). *Touch exhibitions in the United Kingdom, from Museums without Barriers.* London: Routledge.

Perkins Institute for the Blind. (1906). *Course of manual training: pursued at the Perkins Institution and Massachusetts School for the Blind.* Boston: The Perkins Institute for the Blind.

Pollard, S. (2004). *David Blunkett.* London: Hodder & Stoughton.

Porter, R. (1990). *English society in the eighteenth century.* London: Penguin.

Porter, R. (1997). *The greatest benefit to mankind: Medical history of humanity from antiquity to the present.* London: Fontana Press.

RBAS. (1993). *The Royal Blind Asylum and School: 1793–1993: 200 years of service.* Edinburgh: The Royal Blind Asylum and School.

Revesz, G. (1950). *Psychology and art of the blind.* London: Longmans, Green.

RLSB. (1970). *Twenty four selected articles.* (F. J. H., Tooze, Ed.). Worcester: College of Teachers for the Blind.

Sacks, O. (1993, May). *A neurologist's notebook: To see and not to see.* New York: The New Yorker Magazine.

Sacks, O. (2001). *The island of the colour-blind and Cycad Island.* London: Picador.

Sacks, O. (2003, July). *The mind's eye: What the blind see.* New York: The New Yorker Magazine.

Salmon, E. (1889). *How the blind are educated.* London: The Strand Magazine.

Schiff, W., & Foulke, E. (1982). *Tactual perception: a source book.* Cambridge: Cambridge University Press.

Shakespeare, T., & Watson, N. (2001) Making the difference: Disability, politics and recognition. In G. L. Albrecht, K. D. Seelman, & M. Bury (Eds.), *Handbook of disability studies.* London: Sage.

Solity, J. (1992). *Special education.* London: Cassell.

Spellman, W. M. (1997). *John Locke.* Hampshire: Macmillan.

Spence, C., & ICI. (2003). *The ICI report on" the secrets of the senses."* In association with Oxford University. London: ICI.

Stanier, R. Y., Ingraham, J. L., Wheelis, M. L., & Painter, P. R. (1987). *General microbiology* (5th ed.). Basingstoke, Hants: Macmillan Education.

Taylor, Rev. W. (1828). *The diagrams of Euclid geometry (arranged according to Simson edition) in an embossed tangible form, for the use of blind persons, who wish to enter upon the study of that noble science.* York: J. Wostenholme.

Thomas, M. G. (1936). *The first seventy years.* London: National Institute for the Blind.

Thompson, E. P. (2001). *The making of the English working class.* London: Penguin.

Visser, J., & Upton, G. (Eds.). (1993). *Special education in Britain after Warnock.* London: David Fulton.

Vygotsky, L. S. (1994). *The Vygotsky reader.* (R. Van der Veer & J. Valsiner, Eds). Oxford: Blackwell Publishers.

Warnock, H. M., et. al. (1978). *Special educational needs: Report to the Committee of Enquiry into the Education of Handicapped Children and Young People.* London: HMSO.

Warren, D. H. (1994). *Blindness and children: An individual differences approach.* Cambridge: Cambridge University Press.

Weber, M. (2001). *The Protestant ethic and the spirit of capitalism.* (T. Parson, Trans.). London: Routledge Classics.

Whitcher, J. P., Srinivasan, M., & Upadhyay, M. P. (2001) Corneal blindness: A global perspective. *Bulletin of the World Health Organization, 79*(3), 214–221.

WHO. (1980). *International classification of impairments, disabilities and handicaps.* Geneva: World Health Organization.

Witkin, H. A., Oltman, P. K., Chase, J. B., & Friedman, F. (1968). Cognitive patterning in the blind. In J. Hellmuth (Ed.), *Cognitive studies* (Vol. 2). New York: Brunner/Mazel.

Wolf, F. (1992). *125 years of challenging blindness.* London: Royal National Institute for the Blind.

Worcester College. (1879). *The educational report for the year, presented to the Educational Council on the Annual Speech Day.* Worcester: Worcester College for the Blind Sons of Gentlemen.

Worcester College. (1881). *The educational report for the year, presented to the Educational Council on the Annual Speech Day*. Worcester: Worcester College for the Blind Sons of Gentlemen.

Worcester College. (1885). *The educational report for the year, presented to the Educational Council on the Annual Speech Day*. Worcester: Worcester College for the Blind Sons of Gentlemen.

Interviews

Fish interview: Postal interview with John Fish, Deputy Chief Inspector of Schools involved with the Warnock Report and the 1981 Special Education Needs Act. Interview answers supplied in June 2001.

Prentice interview: Postal interview with Lord Reginald Prentice, Former Secretary of State in the Labour Government of Harold Wilson. Interview answers supplied in February 2001.

Warnock interview: Postal interview with Baroness Mary Warnock, Chair of the Committee of Enquiry, HM Government, 1974–1978. Interview answers received in May 2000.

Williams interview: Postal interview with Baroness Shirley Williams, former Secretary of State for Education in the Labour Government of James Callaghan. Answers received in January 2000.

9 781593 119133

Printed and bound by CPI Group (UK) Ltd, Croydon, CR0 4YY

15/09/2026

14962367-0005